1,000+ SUDOKU

EASY
MEDIUM
HARD

PUZZLES

Collin Deloach

Contents

Your mission

is to solve the puzzle by filling in the empty cells with numbers from 1 to 9 without repetition in each row, column, and sub-grid.

The goal is to use logic and deduction to find the missing numbers and complete the puzzle.

						3		2
				8		4	9	1
	1				3			
2			7	4	1	9		6
6			8	9	2			7
9		1	6	3	5			4
			4				8	
8	4	6		7				
7		9						

Without repetition in each sub-grid

						3		2
				8		4	9	1
	1				3	5		
2						9		6
6			8		2	1		7
9		1				8		4
			4			7	8	
8	4	6		7		2		
7		9				6		

Without repetition in each column

Without repetition in each row

						3		2
				8		4	9	1
	1				3			
2						9		6
6			8		2			7
9		1						4
1	2	3	4	5	6	7	8	9
8	4	6		7				
7		9						

Enjoy!

EASY

Puzzles

Easy # 1

7				1				6
		6	5			3		1
		2		3			7	8
3		9		6				2
2	5		9		3		1	7
4				2		9		5
5	2			9		8		
8		4			6	7		
6				8				3

Easy # 2

		8	6	4			7	
	6			5	1	4		
	5		3		2		1	
	7	2	9	6			5	
8								2
	9			1	8	7	4	
	8		5		4		6	
		7	8	2			3	
	2			9	7	5		

Easy # 3

6		3	8			2	1	
9				5	1			6
	8	1		7				
	1						6	2
	4		7		2		5	
8	2						7	
				2		7	9	
2			1	3				4
	3	4			9	5		1

Easy # 4

				6	1			9
2					4	5		
1	7			5	9	6		
	9	3		2				1
8		2	6		7	9		5
6				1		2	7	
		6	4	3			2	7
		1	8					4
3			1	9				

Easy # 5

7							6	2
		1	8	6		4		
	4			2	9	7		
	7	5			4			8
	2	3		8		1	5	
9			5			6	4	
		2	7	9			8	
		7		4	8	5		
5	8							6

Easy # 6

6				2	1	8		
						3		4
9	1		7	8			5	
	5	9			7	4	6	3
1	6	4	5			7	8	
	3			7	6		2	1
5		6						
		1	2	3				8

Easy # 7

	3						2	
					6	1		4
8			2		4		7	9
3			1		8	5	9	
5		9		4		8		7
	8	7	6		9			1
9	7		5		3			6
6		8	9					
	2						8	

Easy # 8

		2						
5			8	1				
	8	3		2	6		4	1
	7		5			4	1	
8	3		2		1		7	5
	6	5			9		2	
6	5		3	4		7	8	
				5	7			4
						1		

Easy # 9

	3	9		5	1		6	
		6						5
				6	8	9		
	9	8			2		7	4
6	4		7		5		9	3
3	5		8			2	1	
		1	5	2				
9						7		
	8		9	7		6	5	

Easy # 10

8		1						
		6	9	2				7
	4			6	5		9	3
1	7	8	5			3	4	
	6	5			4	8	7	9
9	2		7	5			1	
6				1	2	9		
						7		4

Easy # 11

	7	6		1		2		
8				4	2	7		
		1			5			4
6					4		9	3
		2	1		9	5		
9	1		8					6
7			9			6		
		8	4	7				2
		4		2		9	7	

Easy # 12

					7			1
2		6		9	1			
	9		3	6		8		7
9	3	8		7				
6	2		1		3		9	5
				2		4	3	8
3		1		4	2		5	
			5	1		3		2
7			9					

Easy # 13

4				7				
3		8	6		2		4	
			3		8		9	
		9			6		3	1
1		6		2		9		4
8	4		1			2		
	8		2		5			
	7		8		9	5		6
				1				9

Easy # 14

5				8				4
	8			7		1		2
	3		4				9	8
2	7			1				9
3		6	5		7	2		1
1				4			7	5
6	5				2		4	
8		3		5			1	
4				6				3

Easy # 15

9		5	7		6	3		2
8			4		2			1
				8				
	6		2		4	8	1	
		2		6		7		
	9	4	8		1		2	
				9				
7			3		8			6
2		8	6		7	5		3

Easy # 16

	7		9				3	
4			5			1	2	
		8		6	4		5	7
3	8					6		9
		4				8		
6		7					1	3
5	4		3	8		7		
	2	9			7			1
	6				1		8	

Easy # 17

2		7	8					5
	3	8		7	9	6		
	9		4	5				
	6	3	5		8	1		
				3				
		2	9		6	4	8	
				8	7		5	
		5	1	4		8	6	
1					5	7		2

Easy # 18

					5		1	
5	6	4						7
2				4	8			9
	8				3	7		2
3		6	8		7	1		5
1		9	2				6	
7			5	2				8
6						5	2	1
	5		3					

Easy # 19

	9				8	7		
1		3			5			
4	7						9	3
		9			1		7	5
2	6		5		4		1	8
5	4		8			2		
7	5						3	2
			4			9		7
		6	7				8	

Easy # 20

7	5							9
		3		4	8		1	
	2	1		7				3
6	1		2	5				
3		5		9		4		1
				8	1		9	5
5				3		1	8	
	4		9	6		3		
2							5	4

Easy # 21

			4	7	3	2		
		7					6	
	2		6		9	7	1	
7		6	9			8		2
		4	8		1	5		
2		8			7	4		1
	9	5	7		2		4	
	7					9		
		1	5	9	8			

Easy # 22

		9		4	5	8		
6	4							1
		1	3	6			9	
5			9			2	1	
	2	8		5		7	6	
	9	4			2			3
	5			3	1	6		
4							5	2
		2	5	9		1		

Easy # 23

8			5		7		2	
			9			8		
	2	6		3		7		9
5	6		2	9				
	8	3				9	6	
				5	6		3	1
7		5		1		3	8	
		8			5			
	3		7		4			6

Easy # 24

2	5		1	3		8		6
			5		6	2	7	
8	1						9	
6	8		4			7		
		5				6		
		1			7		8	5
	6						3	7
	3	2	6		4			
5		4		7	8		6	2

Easy # 25

	7	8		6	3		9	
		4	7					8
6	9			5	8		4	
				2	5	3		6
8		2	6	3				
	3		2	1			8	4
1					4	9		
	8		3	7		5	2	

Easy # 26

8		4			9		1	
						7		9
7		9		2	6	8		
			2	9		6	3	
9			8	1	3			5
	3	1		6	4			
		5	6	3		9		2
6		7						
	4		9			5		7

Easy # 27

1	4				6	9	7	5
	3	7	9	1				
5				4				3
	8		4	5		7		
		3				4		
		4		9	2		3	
9				2				7
				7	1	5	4	
4	7	5	8				2	1

Easy # 28

	3	7	5	9	8		2	
	9				3	6	5	
	1		2					8
			7	6			4	2
	7						8	
9	4			3	2			
1					5		7	
	5	3	6				1	
	6		9	4	1	2	3	

Easy # 29

		2	5					
7			3	4				9
6						1	4	5
9	6		8			3		
5	2		6		3		1	8
		1			9		7	2
2	5	9						1
3				9	5			6
					8	5		

Easy # 30

6		2				1		
8		4	9				5	
	5		6					9
5	6			7			1	2
		8	3		1	4		
4	1			9			3	5
7					5		8	
	4				3	5		7
		5				3		1

Easy # 31

	8	2	3			1		
1			7	2	9	5	8	6
				5			2	
		6		1				
8	3		9		6		1	5
				3		7		
	9			7				
6	7	8	4	9	1			2
		4			3	6	9	

Easy # 32

	8			2	4			6
	9		5		7		4	
7			6	1			2	
6	5		4	3			1	
		2				4		
	7			9	1		6	2
	6			5	9			4
	3		2		8		7	
5			3	7			9	

Easy # 33

8						2		
					1		6	
2	6	1		8	4	5		7
5	3		8			7		1
1			7		2			4
9		2			3		5	6
3		5	4	9		6	8	2
	9		3					
		7						3

Easy # 34

	8	6	4			2		
				5			8	
3			7	8	2	9	5	6
		5		4				
1	7		6		8		4	9
				7		6		
6	9	1	8	3	5			7
	3			1				
		7			4	3	9	

Easy # 35

	7		8	6			3	
		2		5				
	8		7	3	4		2	5
2				7			8	
8	1						9	7
	6			9				2
7	9		6	8	5		1	
				4		6		
	4			1	3		7	

Easy # 36

	4		8	2		5		
9			3	6			2	8
		8			1	3		
5					3		8	
7	2			5			1	4
	1		4					5
		2	5			7		
4	8			9	7			3
		7		3	4		5	

Easy # 37

2		4	9	6			1	5
			1		4	3	2	
6		3				4		
5		9	6				7	
	4						5	
	6				1	9		4
		8				7		9
	2	6	4		5			
4	9			3	7	5		2

Easy # 38

				7	1	3		4
	1	4	6			7		
	5	7	3				2	1
9		3	1					7
		1				5		
7					9	8		6
5	7				3	6	8	
		6			7	4	3	
4		9	8	6				

Easy # 39

7						3		1
1			5	7	3	8		
	5		8				9	2
	3			4				8
	1	5				4	6	
9				5			3	
3	7				5		2	
		9	4	6	2			3
6		4						5

Easy # 40

9				3				2
4	3				1	8	5	9
	2	5	8	4				
	6		3	9		5		
		2				3		
		3		8	7		2	
				5	4	9	3	
3	5	9	6				7	4
8				7				5

Easy # 41

	9			5			6	
6	4			2		3		
	5	2			1	9		
	3			9		7	2	
5	6		2		7		3	1
	1	7		3			8	
		6	9			8	4	
		4		7			1	3
	2			4			9	

Easy # 42

	5					3	9	
8	1		4		2			
3		4		7	9	8	2	
1					6	4	3	
4								2
	3	2	1					9
	4	8	3	1		2		6
			6		4		7	8
	7	1					4	

Easy # 43

	7						4	
8		6	3					
5	1		8		7			2
	5	9	2		6			4
1		2		8		5		9
6			5		3	1	2	
3			4		9		1	5
					5	2		3
	2						7	

Easy # 44

				1			5	7
9	7				5	4		8
		4	7	3		6		
	4	9					7	
	3		9		1		2	
	1					5	9	
		2		8	7	9		
3		7	6				8	2
1	6			9				

Easy # 45

	8		1	6			7	
			4		3			
5	6						4	8
2	3		8		6		5	1
		4		3		8		
8	5		9		1		3	6
3	2						6	9
			6		9			
	7			5	8		2	

Easy # 46

		4			5		3	
6	3		8	7			4	1
			2	3			9	
4		7	1	5				
1								9
				6	7	4		5
	4			8	6			
5	8			4	1		6	2
	1		3			8		

Easy # 47

	4		2	8	6		5	3
		2			4		7	
1	6		5				8	
	9	4		1	3			
	2						3	
			4	5		8	9	
	7				1		6	5
	3		6			7		
4	5		7	9	8		1	

Easy # 48

		7		5	6		3	
		5	8		2	6		
	1		7	3		4		
	2	4	9	7		5		
1								2
		9		6	1	3	4	
		2		9	4		5	
		1	5		3	7		
	4		1	2		8		

Easy # 49

	4	3	9		1			5
	7	9		2			4	
8			7		4	9		
9		7		4				
			5		3			
				7		5		6
		2	4		6			9
	9			8		2	1	
3			2		9	8	6	

Easy # 50

1		7						4
	5			3	9	1		
	6	2		5				7
7		9	6	2				
6	1			9			7	5
				7	4	6		3
5				8		4	6	
		6	2	1			5	
9						7		8

Easy # 51

	4	5	9		3		6	
7		1	4		6		3	
		6						
		2	1			6		7
1	6		8		5		9	4
4		8			7	3		
						1		
	1		3		4	8		2
	2		7		1	9	4	

Easy # 52

				6				
	7	6	9		3	4		5
3					7		6	
6		4			9	2	5	
7	9		8		5		1	4
	3	1	6			7		8
	4		3					1
2		5	1		4	8	3	
				7				

Easy # 53

		8				2		9
		1		2			7	4
7			5	3			1	
				9	4	6		7
	3	7		8		1	9	
8		9	7	5				
	1			6	8			3
5	7			1		9		
9		3				4		

Easy # 54

		4	7	6				
7			8			2		
3			2	4		5	9	
		3		7			5	9
9		8	3		5	1		6
4	6			9		7		
	5	7		1	6			3
		9			2			1
				3	7	6		

Easy # 55

	3		5	7		2	4	
	1	5						
8			4			1	3	
7	9			5	8			
		4	6	9	7	3		
			2	4			5	7
	8	6			4			9
						4	1	
	4	1		2	5		6	

Easy # 56

		1				5		2
	9				7	4		3
7					2		9	
5	1			6			2	9
		4	1		8	3		
9	8			7			1	4
	3		9					6
6		9	8				4	
1		8				9		

Easy # 57

			3			9		
	3	5	9				8	4
		6			4		5	3
6						7	2	5
7		9				4		8
5	4	3						9
8	9		1			5		
2	5				9	3	1	
		1			2			

Easy # 58

				1	4	7	6	
4			8		5	2		1
	7					9	5	
	5	4						9
			3	6	8			
6						3	1	
	3	1					2	
9		5	2		3			6
	6	2	5	4				

Easy # 59

	8	7		1			3	9
3			7		4	2		
	9				8			
				8	2	1		3
9	3						5	2
2		8	6	5				
			5				9	
		9	8		7			6
6	2			3		5	7	

Easy # 60

	4			8			3	
	8	5			4	2		
6	9			7		8		
	5			6		7	9	
9	6		7		3		2	1
	3	7		4			6	
		6		3			8	2
		4	9			3	1	
	2			1			4	

Easy # 61

				1	5	2		6
2	8		3					1
1	5		8			7	3	
5		1	6			3		
7								4
		3			4	6		8
	9	4			8		7	3
3					1		4	2
8		2	4	3				

Easy # 62

	7	5	4	2	6			
	4	2	8			1		
6			9	5		7	4	
	5					6		8
7		6					1	
	3	4		6	7			1
		7			5	4	6	
			3	9	4	8	2	

Easy # 63

	1			7		3		6
8	3		9			7		1
		5	4					
9		1		4	2			
6			1	5	9			4
			6	8		1		9
					4	2		
1		2			5		3	7
3		8		1			9	

Easy # 64

	8		9	7			2	
	7		4	3	5	8	6	
5				2				
		1		6			5	
	6	8				3	7	
	3			8		1		
				4				1
	1	4	2	9	8		3	
	9			5	3		8	

Easy # 65

1	3	8					5	
	5				9		1	
9	2	4		1	5			8
8			2	5				
			4		7			
				6	3			5
6			9	4		5	3	1
	4		1				7	
	1					6	4	9

Easy # 66

	1	6		4				
9			7		6		3	
	7			2	9	4		6
		5	4			6	9	
	8			9			2	
	3	9			1	8		
4		1	3	8			6	
	9		6		5			4
				7		5	1	

Easy # 67

4		3			7	9	5	
7					5	6		
9	6	5		1	4			
	4					5		
1	8	2				3	6	7
		6					4	
			7	2		1	9	5
		7	3					6
	9	1	5			7		8

Easy # 68

							2	
	2		9	5	1	4		
4	6			3	2	8		
		4			3	9		7
	8	2	4		6	5	1	
7		1	5			2		
		3	6	2			4	5
		7	3	4	8		9	
	4							

Easy # 69

		1	5	2			6	7
		7	9	3	8		5	
	5							
4		8	2			7		
	9	3	6		7	5	1	
		5			3	9		4
							7	
	8		1	7	2	4		
3	7			5	6	2		

Easy # 70

	3	5			1		2	6
2					5			
	1	8	2					3
6	7	3				1		
1		4				8		7
		9				3	5	4
9					7	6	3	
			6					1
3	6		1			7	8	

Easy # 71

		1		8		4		
9	4	2	3			6	8	
				4	6		2	9
2				1	8		7	
7								2
	3		2	9				4
4	7		1	6				
	2	6			5	9	4	1
		9		2		7		

Easy # 72

6			2		3		5	
	2	7	5	8				3
				7		2		9
5		2			7	4		
8				5				1
		1	9			5		6
9		4		3				
2				1	6	9	7	
	7		4		2			5

Easy # 73

	3	4			2	8		5
					7		1	
	5	9		3				4
			6	7		2	4	
		7	2	1	4	9		
	4	2		8	9			
2				4		5	8	
	6		7					
5		3	1			4	6	

Easy # 74

9	8			3		7	4	
	4				6			
3	1				8	2		5
				6	5	4	1	
	6		8		9		5	
	2	7	3	4				
1		9	6				2	3
			5				6	
	5	4		9			7	1

Easy # 75

	8		7					2
	3				4	6	5	
	4	9	5	3	2		7	
			9	6			1	7
	9						2	
3	1			4	7			
	6		3	1	8	7	4	
	5	4	6				8	
8					5		9	

Easy # 76

		7			8	4		
8	2	3		4	7			1
4	1	6				7		
1			3	7				
			2		5			
				9	6			7
		4				2	9	8
9			8	2		6	7	4
		2	4			5		

Easy # 77

8				7				2
		2	1			4		7
		5		4			8	9
4		3		2				5
5	1		3		4		7	8
6				5		3		1
1	5			3		9		
9		6			2	8		
2				9				4

Easy # 78

	8	3	6	7		4		9
				8	4		5	
								7
3		8			5	1		
	5	1	8		7	9	4	
		7	2			6		5
8								
	3		1	5				
1		4		3	9	5	6	

Easy # 79

			8				2	4
4		5				1		9
	6		4			3		
5		8	3				9	
9		6	5		8	7		3
	2				7	4		5
		2			3		4	
8		4				2		1
7	1				5			

Easy # 80

						9	7	
	9	7		2	3	8		
	8	4			7			6
			2	7		3		1
	7		8	6	1		5	
1		6		3	4			
4			7			5	9	
		5	3	1		7	2	
	3	9						

Easy # 81

	5		9	6		1		
		7	2		5	8		
		3		1	8		9	
	9	2	8	4		6		
1								8
		5		7	6	9	1	
	2		4	5		7		
		4	1		3	5		
		9		2	7		8	

Easy # 82

4					6	7		
3	2			4				9
		5		7	9			2
		3			7	8	1	
9			4		1			6
	4	1	5			3		
5			7	2		9		
7				9			2	1
		2	1					3

Easy # 83

1		8	6	9		7		2
	2				3			6
			1	2				9
	9	7	5	1				
		4				6		
				7	6	9	5	
4				3	8			
3			7				9	
9		6		5	2	1		3

Easy # 84

	9			8	1	6		
8			6			7	2	5
		6	5		4		9	
6		9					3	
	8						1	
	5					2		4
	2		1		5	3		
1	3	7			8			9
		4	9	3			8	

Easy # 85

			5			9	6	
	4		9					3
7		9				8		1
5		7	3				8	
4		8	7		5	3		2
	6				2	7		9
9		5				1		6
6					3		9	
	1	2			7			

Easy # 86

			7		4			
	6	5				7	4	
		3		2	8	5		
	8	2	4		1	6	7	
9				6				8
	5	6	8		7	2	1	
		8	1	7		3		
	2	7				9	8	
			9		6			

Easy # 87

		3	7				9	
4		2	6			3	5	
			3	1		2	6	4
		9						8
5	2	1				7	3	9
8						6		
9	4	6		2	8			
	8	7			3	4		6
	3				6	9		

Easy # 88

				4				
	9				5	2		
6		5	2		9		7	1
4	6				8	5		9
	2	9	1		6	3	4	
7		1	3				8	2
2	1		5		3	4		8
		8	4				5	
				8				

Easy # 89

	5	1	9					6
		3	1	6		9	2	
2	6			3				
			6					5
7	2		5	9	1		4	3
9					8			
				1			6	2
	1	8		2	7	4		
3					6	1	8	

Easy # 90

		6				8	1	4
		9	4	1		5		
	4		2					
3		8	1				6	
6		2	5		9	4		8
	5				2	1		9
					4		8	
		1		7	5	3		
7	6	4				9		

Easy # 91

2	8						1	4
	4				2	3	5	
	9		4	3		8		
	6				5	1		
3		1	8		9	4		5
		9	3				7	
		4		2	3		8	
	3	2	7				4	
8	1						3	9

Easy # 92

7	8			9			1	
3	1		7		2	6		
		5	8		1			7
8		7		1				
			6		3			
				8		4		6
9			1		4	7		
		3	9		7		4	5
	7			5			2	9

Easy # 93

3		5		2			6	9
7		2			3	8	4	
9					1			
				1	8		9	7
1			3		5			8
4	6		2	9				
			8					1
	5	7	1			2		4
8	9			5		7		6

Easy # 94

			5			8		4
8	2						1	3
		1	7				2	
5	1		4			2		
7	4		3		5		6	9
		9			7		3	5
	7				1	6		
9	8						5	1
1		2			3			

Easy # 95

			5	3				2
5	8		9	2			1	3
		3			7			9
	1	2	4	5				
	6						9	
				1	9	4	2	
7			1			2		
2	9			4	3		5	7
6				7	8			

Easy # 96

	8	2						
4			2			9	3	
	9		7	1		8	2	
6	7			2	1			
		5	6	4	9	2		
			3	7			4	6
	2	1		6	7		5	
	5	8			2			3
						7	8	

Easy # 97

	1	7			6	4		
6			3					1
4	9	3			8			
7	5			2				8
	6			1			5	
2				8			7	4
			5			6	4	7
5					1			3
		9	7			8	1	

Easy # 98

	4						8	6
				1	3		5	4
		3	2		8	1		9
3	8					6		
			7	5	2			
		5					1	7
8		6	9		7	5		
9	5		8	3				
1	7						9	

Easy # 99

1			8		7	9		
9		3		4	6		7	
	7					5		
	3		5			6	1	
5		2		7		4		3
	1	7			9		8	
		1					5	
	6		7	2		1		9
		5	1		4			8

Easy # 100

1		2	4		5	9		8
				6				
		5	8		2	4		
	9		3		8		6	7
4				5				9
8	3		7		9		5	
		3	9		7	8		
				8				
2		9	5		4	6		1

Easy # 101

		5		9	6	1		
			7		8			
	1	8				9	2	
	6	2	9		1	7	3	
1				7				8
	9	7	6		4	2	1	
	4	9				3	7	
			4		9			
		3	1	2		5		

Easy # 102

		6			2	3	5	
		7	6	5			4	
2		4				8		6
		9			3		8	
5	8		4		7		6	3
	7		5			1		
4		8				5		7
	6			2	5	4		
	2	5	1			6		

Easy # 103

		7			6			
1		8		3			2	
			2	7	1	4		8
4				9	5	2		3
8		2				5		7
5		9	7	8				1
6		5	3	2	8			
	8			4		9		2
			6			8		

Easy # 104

	4	6			3		2	5
	9	3		4		8		1
		8			7			
				7	2	6		8
		7	3		9	2		
1		5	4	8				
			2			7		
8		2		9		1	6	
9	6		7			5	4	

Easy # 105

6		8	3		1			4
		2	8	6	9			
3						6		
	1	9			3	4	2	
		4	9		2	8		
	3	7	6			9	1	
		3						7
			4	3	5	1		
1			7		6	3		2

Easy # 106

			3	2	4		7	9
4	9			7	6	3		
	8				1		2	4
	3	1						7
8						9	3	
3	4		7				9	
		8	9	3			4	5
2	1		4	6	5			

Easy # 107

		2	4	5		9	8	
	9	8	3					7
5	3			8				
			9					6
7	2		8	6	1		5	4
1					3			
				7			3	5
3					6	8	1	
	5	6		3	8	7		

Easy # 108

2		3		6	9		5	
				5	2			9
4					8		6	
	7	9		4				2
1	4		5		3		9	6
5				2		3	4	
	2		1					8
7			2	9				
	5		8	7		4		3

Easy # 109

	9	4	5	3		2		
	1			7	4			3
							6	8
9		2			5	1	8	6
6	4	1	2			3		5
1	2							
4			7	8			3	
		8		5	1	7	4	

Easy # 110

	7		1	3		2	9	
1					4			8
		4	9	7				1
	4				9	5		
	9	5		4		8	1	
		2	7				4	
4				8	2	9		
7			5					2
	2	8		6	7		3	

Easy # 111

	4				5			2
5		7	6					4
6			1	9	4	8		7
9	1			7	8			
2								3
			2	6			8	9
7		2	5	1	3			8
1					7	6		5
4			8				3	

Easy # 112

9	4		2				5	
		6			7		8	
	7		6	5	4		2	1
	3	7		9	1			
	6						1	
			7	2		5	3	
7	2		8	3	5		9	
	1		4			8		
	8				9		4	2

Easy # 113

	8	2	3				5	9
6		4		5			1	3
			7					6
9		6	8	7				
8			1		3			7
				6	5	4		2
7					8			
4	9			1		6		8
2	5				7	1	9	

Easy # 114

	4						3	7
	3		2	4	7			6
		2	6			1	9	
		7		5			6	
2		3				8		5
	1			2		7		
	7	4			2	9		
1			5	8	9		7	
5	8						2	

Easy # 115

3							8	5
				4	6		3	2
		6	1		5	4	9	
5	6					8		
			7	2	1			
		2					7	4
	5	8	9		7	2		
2	9		5	6				
7	4							9

Easy # 116

				2			1	
		2		9	4	6		
6		3	1	8	7	9		
		1		3				5
8		9				3		6
5				6		8		
		8	6	4	2	5		7
		6	8	1		4		
	5			7				

Easy # 117

1		3	9				6	2
8	9			7				
		2		6	3	7		
	8					5	7	
	1		7		8		2	
	4	7					3	
		4	3	1		9		
				8			5	3
7	3				5	4		6

Easy # 118

				2	7		8	
		8	1	3	4	6		
	4	6	9					3
				8	9	1		5
9		4				8		6
8		3	5	4				
1					5	3	6	
		5	8	7	3	2		
	3		2	6				

Easy # 119

6			3		4	5	9	
3			1		9			
				8		4		
2		3	8				1	
	5	8		1		2	4	
	4				5	8		7
		2		6				
			7		3			4
	3	7	5		1			2

Easy # 120

	4		2	5		1		3
1		5	3	9	4			
3		9	8			6		
5						4	8	
	1	4						6
		1			5	3		4
			7	2	3	8		9
7		3		4	1		6	

Easy # 121

				1			3	
4		7	3	9	5			6
1				6	8			7
3				4		2		
6		9				7		4
		2		7				9
7			9	3				8
9			7	8	1	5		2
	2			5				

Easy # 122

	1		6			5	3	7
4				1	9	6		
		6	3		2			4
	6	4						8
1								9
3						7	2	
7			9		3	8		
		2	4	8				1
8	9	5			1		4	

Easy # 123

			3	5	4	8		7
5					6			
8		4		2			3	
		7		9	1	2		3
3		8				5		1
9		1	5	8		4		
	8			7		3		9
			6					8
1		6	2	3	8			

Easy # 124

		8				7		4
	5		8			3	9	
		3	5	7	4			1
	3			8		1		
4	7						6	8
		2		4			3	
2			3	9	8	6		
	1	5			2		8	
3		6				9		

Easy # 125

3		2		1	7		5	
9			8	3	5		6	
								3
	1				2	6	4	
4	2		7		3		1	8
	9	6	5				3	
1								
	3		4	2	9			1
	8		1	5		3		7

Easy # 126

8				5	9			
							9	
4		6	1	3		8		5
1						5		8
5	6	3	4		8	1	7	2
7		2						6
3		8		4	5	2		9
	1							
			9	8				7

Easy # 127

7						4	2	5
9			8	4				1
	3		5					
1		7	6				8	
5		3	7		8	2		6
	2				1	9		3
					6		5	
8				1	5			7
3	1	5						2

Easy # 128

4		7	6	8	9		5	
3	5							2
			2		5			
5		4		6	3	8		9
		3				5		
2		6	5	7		1		3
			7		2			
8							9	7
	3		8	4	6	2		5

Easy # 129

5	9	2		6	1			4
1	8						6	
					4		9	5
				7	6	2	3	
7			4		2			8
	2	6	8	3				
4	7		5					
	1						8	7
6			9	2		4	1	3

Easy # 130

		4	8		1		6	
8				2		4		1
	7		9		4	5		8
				8		1	4	
			5		7			
	3	7		1				
3		6	4		2		5	
9		2		6				4
	4		3		8	2		

Easy # 131

5	9		4					
4			2	3		8	9	5
	3						7	2
	6	8	3	1				
7			8		4			1
				6	7	3	8	
1	7						2	
6	2	4		8	9			3
					5		1	4

Easy # 132

		4	1	8		3		
			6		5			
6		5				4		7
5		7	2		6	8		1
	1			7			9	
2		8	5		1	7		4
1		9				5		8
			7		9			
		3		5	2	1		

Easy # 133

5			6					
		2		3			9	4
	1	9	8				3	5
3	2			4	7			
	6		2	8	3		7	
			5	6			2	3
1	3				2	9	4	
9	7			1		3		
					6			8

Easy # 134

2	8	5	6		7			
	4			3	1			
3				2			6	7
5	3		1	9			7	6
1	6			8	5		3	2
4	5			7				3
			3	5			1	
			8		4	7	5	9

Easy # 135

		3			9	2	6	
2	9					7	1	4
						9	5	
6		9		4	7			
		2	9		5	1		
			8	1		6		2
	3	1						
9	7	8					2	1
	2	6	3			5		

Easy # 136

8		6	2		3			
		4	8				3	6
	7		6		4		8	
5			4				1	
7	1						9	5
	6				1			8
	5		9		6		4	
9	8				7	2		
			1		2	7		9

Easy # 137

				2				
	6	5	3		4	8	9	
	4		9		5		3	
2		1	5		7			6
		6		3		4		
3			6		1	5		7
	5		1		6		7	
	2	8	4		3	9	6	
				5				

Easy # 138

	2		8		6			
				9		1		
	5		2		1	3		6
	7	2	9					8
3		9		8		7		1
1					3	9	4	
2		4	3		8		7	
		7		5				
			4		2		1	

Easy # 139

	4	2	3		5	8	6	
				9				
		1	2		7	9		
1	9		7		2			3
	5			3			2	
2			1		9		7	8
		3	9		4	5		
				8				
	6	4	5		3	2	9	

Easy # 140

5	3							
		9		7	6	2	1	
7			2	8			4	
3	5	4	6			9		1
6		7			9	4	2	3
	7			5	8			2
	2	8	4	6		5		
							9	4

Easy # 141

9			1	2		7	4	3
		2				5		1
3		7	9					
	4	8	2	6				
5			4		9			6
				8	5	4	2	
					3	6		9
6		5				1		
8	9	1		4	7			2

Easy # 142

	1		2	8		9	5	7
8							2	4
7	5		1					
3		9	8	6				
	4		9		1		6	
				3	4	8		9
					5		1	6
4	6							2
2	3	1		9	7		8	

Easy # 143

3								1
6	5		8		7		4	
	4	1	5					
1		6	4		5		9	
	8	5		2		1	6	
	7		9		1	8		5
					4	9	2	
	1		3		2		5	6
7								3

Easy # 144

	8	1	5		7		6	
7		3						5
6		5	1	4				
	6					7		3
			7	6	9			
1		4					8	
				3	4	2		6
2						8		1
	4		9		1	5	3	

Easy # 145

		2	3	4		7	8	
	8		6	2				
4							2	
1		5	9			8	6	
7		8	4		5	1		2
	9	3			6	4		7
	5							8
				9	4		3	
	2	4		5	8	6		

Easy # 146

	4			9	6			
	1		2					9
	9	7		4	1	3	6	
				6	8	7		4
		1				5		
8		4	1	7				
	2	6	9	8		1	4	
4					7		2	
			3	2			5	

Easy # 147

7			8	9		4	3	
9			2			1		
		8	4	7				
		4		1			8	5
8		9	3		7	6		1
1	3			4		7		
				8	4	5		
		2			6			4
	1	3		5	2			7

Easy # 148

1	7			9				
5		9	1				8	
		2	4	7		5		9
			5				6	
2	8		9	6	3		4	7
	3				1			
7		6		1	9	8		
	1				6	9		3
				8			7	1

Easy # 149

	3	9	2		5			8
	2			1				
			9		4			5
		4			1		5	7
	7	2		4		3	1	
6	1		3			2		
2			5		6			
				8			7	
7			4		3	5	6	

Easy # 150

				9				6
6	2		5				3	
		7	1	6	3	2	8	9
	9			5				
1		4	2		6	8		5
				1			2	
8	4	2	6	7	9	1		
	1				5		7	8
7				4				

Easy # 151

4			9					
		6				2	4	7
		3	4	7		1		
	5	2	7					6
	6	9	1		3	4	2	
1					9	7	3	
		7		8	1	5		
6	8	4				3		
					4			2

Easy # 152

2	5	8				4		6
	4	7						
	6	3	4				9	
			5	2			4	3
	8		7		4		6	
6	3			8	1			
	7				9	6	3	
						9	8	
8		6				5	1	4

Easy # 153

3			5	1			4	
4		9				5		8
5					9		1	2
6					2		8	
	8	1	4		3	2	5	
	3		1					7
1	9		7					5
8		4				3		1
	5			9	1			4

Easy # 154

8		7		2	3		4	6
		5						
	2		7	6				
9			1			6		4
7	6		5		9		8	3
5		2			6			7
				5	8		6	
						9		
2	5		4	9		3		8

Easy # 155

5		3		1			6	8
			6					7
	7		9		8	3		
	9	5	3	6				
1		7				5		6
				9	5	1	4	
		1	8		2		5	
7					9			
9	8			4		7		1

Easy # 156

	9			3	2			5
7			5	6			3	
	1		4		7		2	
5	4		2	8			6	
		3				2		
	7			1	6		5	3
	8		3		9		7	
	5			4	1			2
4			8	7			1	

Easy # 157

				7				
6	2		3		1		4	5
	9		2		8		7	
7		9	8		2	3		
1				3				2
		2	9		7	4		8
	3		7		6		1	
5	6		1		3		2	7
				4				

Easy # 158

	7	6	3		2			9
	1	2		6			3	
3			7		4	2		
7		5		8				
			9		5			
				4		8		3
		3	4		8			6
	4			2		3	8	
5			1		3	9	4	

Easy # 159

5	3				8	9		
			5		7	3	8	
8			9		3			1
2					9		6	
4	6						1	2
	8		2					3
9			3		4			6
	4	1	7		2			
		7	1				4	8

Easy # 160

6				9	5		7	
	1		3	4				
	8		6	7			2	3
		9						1
	4	3	1	5	9	2	6	
1						3		
7	6			3	4		1	
				1	7		8	
	5		8	2				9

Easy # 161

	9	1	3		5			8
			9		4			5
		3		2				
	4				2	5		7
	3	7		4		2	1	
6		2	1				3	
				8		7		
3			5		6			
7			4		1	6	5	

Easy # 162

			2	3		6	8	5
		5	6				7	
6		8	7			1	2	
		6						2
5	7	1				4	3	9
2						5		
	9	7			6	3		8
	5				1	7		
8	6	3		4	7			

Easy # 163

						3		
2	1		6	8			9	5
7				9	2			
4							1	7
1	3	7	9		8	4	5	6
9	6							3
			2	6				9
6	9			5	3		4	8
		2						

Easy # 164

	1	9				6	7	
			7		6			
	4			8	2		1	
	8	2	6		3	7	9	
5				9				2
	9	1	2		7	3	8	
	2		3	7			4	
			5		9			
	7	8				2	5	

Easy # 165

	4	9				7	6	
			6		3			
	7		1	4			2	
	3	8	7		4	1	9	
6				3				7
	9	7	5		1	4	3	
	2			9	7		8	
			4		5			
	8	3				5	4	

Easy # 166

3		5		6	4	2	8	
2	9		5		8			
	1					3	4	
9					7	5	3	
5								8
	3	8	9					4
	6	9					5	
			7		5		6	2
	5	2	3	9		8		7

Easy # 167

7			3				6	
2		3	6				4	5
			4	1		7	2	3
3						4		
5	6	7				9	1	8
		4						7
1	3	2		8	6			
6	9				3	2		1
	7				5			6

Easy # 168

				1		5		4
8		4			5		9	2
	9		4	7			3	
	8	9				4		
		7	8		1	6		
		1				8	5	
	6			2	4		8	
7	4		3			2		6
1		3		8				

Easy # 169

		7				3	9	
		9		2			4	6
	3		5	8				2
				4	6	9	5	
9		2		5		6		3
	6	8	7	9				
2				3	4		6	
6	7			1		2		
	9	1				5		

Easy # 170

		2	6	8	4		3	5
			7		5			
	6					1	7	
4	5		3	7			2	9
2								3
8	3			4	2		1	6
	2	3					5	
			5		3			
7	8		4	6	1	3		

Easy # 171

	5	6	2	4	8	9		
		1	3					2
		2			9	3	6	
			5	6		4		8
		7				1		
5		4		9	1			
	9	3	6			8		
7					5	2		
		5	7	8	3	6	1	

Easy # 172

		8	9	6		3	4	7
	4	3	8					
	6					9	5	
7	2		6	1				
		5	7		8	1		
				2	5		7	6
	5	1					9	
					3	8	1	
8	9	2		7	4	6		

Easy # 173

								1
		1	8		7	6		3
		8	1		6		2	5
1	2				5			4
	6	7	3		9	1	5	
8			2				6	9
9	4		6		8	5		
7		6	5		2	4		
5								

Easy # 174

5	3		6		2	9		7
		7	9				6	
				7				
4		3	2				7	5
	5	8	3		1	2	9	
9	1				7	6		8
				9				
	8				6	5		
1		6	5		8		4	3

Easy # 175

6			1	5		4		8
		2	8	6			1	
	1				2		7	
2					8	9		
8		9		2		7		1
		4	6					2
	6		9				4	
	2			7	4	8		
4		7		3	6			5

Easy # 176

		3			1	8		5
5		6		4	2		7	
			7	8	5	4		6
	1	7						4
3						7	6	
8		1	5	2	9			
	3		6	7		5		9
7		5	4			6		

Easy # 177

		6	1			7		
5	1				2			3
4	9	2			7			
	2	9		5		8		
	7			1			4	
		5		8		2	7	
			5			9	3	6
9			4				1	2
		1			6	4		

Easy # 178

7	3			1		8		
9					4			
			8	9	3		7	2
	2			5	6		1	8
8	7						9	6
5	6		9	7			3	
6	4		1	8	7			
			4					7
		7		2			8	5

Easy # 179

8	9	3				6	1	
6	4							
7	1		4					2
			8	6		1		7
1			3		2			6
3		7		5	9			
4					3		7	1
							2	3
	3	1				5	6	9

Easy # 180

		9			2		6	4
	2				5	9		
3							5	7
	7	3		8		5	9	
4			3		1			6
	9	1		2		3	4	
1	3							9
		6	9				8	
9	8		1			4		

Easy # 181

9		7	1	4		2		3
			8	7		6		
	2				5	7		
2	4		3	5				
3								6
				9	4		2	5
		3	7				1	
		2		1	9			
5		1		2	3	9		8

Easy # 182

				2	1			
1			3				5	
2	8			6		7		3
	9			5		8		4
8	5	4		9		3	2	1
6		1		3			7	
7		3		4			9	6
	6				3			8
			5	1				

Easy # 183

5	4			7		8	3	
			4				1	
7		8	1			6	2	
3	2		6	5				
	1		9		7		4	
				1	4		8	5
	8	6			9	4		2
	5				1			
	9	7		6			5	3

Easy # 184

3					4	5		
		4	3	2	9	6	8	
	7	9	6			2		
4		1		7	8			
		3				8		
			4	6		1		2
		5			7	9	6	
	4	6	5	1	2	7		
		8	9					5

Easy # 185

5			4	9	8			7
		9			1		4	5
	7		2	3				
8		6	1	7				
7		5				1		4
				4	6	7		9
				5	3		9	
9	5		6			8		
3			9	2	7			6

Easy # 186

	4			1	5	3		9
5			4		9		7	
	2	9		3				
		8	3			9	5	
	6			5			1	
	7	5			2	6		
				4		8	2	
	5		9		8			3
3		2	7	6			9	

Easy # 187

	4	3			7	9		
7	8	6			1			
5			4					1
8	7			3				2
	1			4			6	
3				2			1	7
4					5			6
			3			5	9	8
		8	6			7	4	

Easy # 188

7	1			5		4		
6				4				9
4		2			6	8		
2				1		5		7
1	7		5		9		3	8
9		5		6				1
		6	7			9		3
8				3				6
		1		9			8	4

Easy # 189

		7				8		
2			3		6	1		5
					5		7	2
9			5		2	7	1	
1	7			4			5	6
	6	5	7		9			3
4	9		2					
5		1	4		8			7
		8				3		

Easy # 190

	7				4		2	
4	6	9		2	7			3
2	5	3					7	
3			6	7				
			9		8			
				1	5			7
	2					1	9	4
1			4	9		7	5	2
	9		2				8	

Easy # 191

4		8		1		6	9	
			8			2		
1	9		2			5	7	
6		5	7	4				
		2	3		1	8		
				2	8	9		4
	7	9			3		8	5
		4			2			
	1	3		7		4		6

Easy # 192

		1	4	2	7	3		
	6	4	3					5
				6	1		4	
				8	3	4		7
6		7				8		9
3		5	9	7				
	7		2	1				
4					9	6	8	
		6	8	4	5	7		

Easy # 193

	7		3	8				4
		9		7	5	3		2
1	6							
6	1	4	5			9	2	
	5	7			9	4	1	3
							4	9
3		8	4	5		6		
7				6	8		3	

Easy # 194

7	3	9				1	6	
6		4	1					5
1		2						
			7	3			4	1
9			2		1			6
4	6			9	8			
						5		9
2					5	6		4
	9	6				7	1	8

Easy # 195

			9	1		5	3	
				3	7	8		
3		1	6					
4		7	5			3		8
	2			7			9	
6		5			9	4		7
					1	2		4
		4	3	5				
	1	6		9	4			

Easy # 196

6		2	4		3		9	1
	3				2	6		
				6				
1	6				4	9		8
	2	4	5		9	7	1	
7		3	6				5	2
				2				
		1	3				7	
9	8		7		1	3		5

Easy # 197

9								
1	2		4		9	6		
8		4	3		6	9		
7			1				2	9
	1	9	5		8	3	4	
5	4				2			6
		7	2		1	4		3
		1	6		4		7	5
								1

Easy # 198

		2			5	8		3
5			2	6			9	7
				7	3		5	
8			4		9		7	6
				1				
1	9		5		7			2
	4		6	5				
7	1			3	4			9
3		8	7			5		

Easy # 199

	3		6	9		8	4	
	2	6						
7			4			2	3	
9	5			6	7			
		4	1	5	9	3		
			8	4			6	9
	7	1			4			5
						4	2	
	4	2		8	6		1	

Easy # 200

9	3			4				
8		5	3				1	6
		6		1	5	4		
	9					7	4	
	8		4		9		6	
	2	4					5	
		2	5	8		3		
4	5				7	2		1
				9			7	5

Easy # 201

	1	5	9		3	8	4	
		3	4		5	9		
				2				
8			6		4		7	2
	9			3			8	
6	4		7		8			3
				4				
		6	8		7	4		
	5	8	3		9	2	1	

Easy # 202

4	6							7
9	3	2		7	6		5	
					5		3	9
				8	7	2		1
	8		5		2		4	
2		7	4	1				
8	5		3					
	7		9	2		5	1	6
6							8	4

Easy # 203

	1	7				4	9	
4		2			5			
	9				8			1
9					2	5	1	
	6	3	5		7	8	2	
	7	5	8					3
6			1				8	
			7			1		9
	5	1				3	4	

Easy # 204

3				6		1	4	
5			2			9		6
		4		9		5		
8		6		5		3		
	2	3	8		6	4	9	
		7		3		2		8
		5		1		6		
7		1			5			4
	3	2		8				1

Easy # 205

	1	5		7				
		3		8	2		5	6
6			1		3	4		
	8		5			2	4	
		9		4		8		
	3	4			6		1	
		2	3		7			4
3	6		4	9		7		
				6		5	3	

Easy # 206

6	7	2		8	4			5
4		9				8		
					5	2		6
				1	8	3	7	
1			5		7			9
	8	7	9	3				
5		1	6					
		4				9		1
8			2	7		4	5	3

Easy # 207

	7	9				5	6	
		6	1	9		2		
			5		4			
	3	4	6		9	7	1	
5				4				6
	6	7	8		1	4	9	
			9		8			
		2		7	6	3		
	4	3				9	8	

Easy # 208

	2	4	7				8	
				5		2		
8			3	2	1	4	5	6
	6			8				
4		7	1		6	8		5
				7			3	
6	4	3	9	1	8			2
		1		3				
	9				7	1	6	

Easy # 209

1			2		4			9
				1				
8	6		5		3		7	4
		3	4		2	9	1	
	4			3			5	
	2	8	1		9	4		
4	1		3		5		6	7
				8				
5			7		1			3

Easy # 210

9			7		2	6		
	7	8		1			4	9
	6				8			
				8	9	2		4
4	8						1	6
1		5	4	2				
			2				6	
6	1			5		7	2	
		4	3		7			1

Easy # 211

		5			3	1		
4	3	7		1	5		2	
2	1	8				5		
	2		7	5				
			4		9			
				6	8		5	
		1				4	3	6
	6		3	4		8	1	5
		4	1			9		

Easy # 212

		8	6			2		
		6				8	9	5
	5		9	8		4	6	3
				5	4		3	
			8		2			
	1		7	3				
8	9	7		6	3		1	
1	6	4				3		
		3			9	6		

Easy # 213

4		9	2	8	7			1
5			1				7	
8					4	6		2
			9	6			1	3
9								7
3	8			4	1			
2		4	6					5
	5				2			9
6			8	3	5	1		4

Easy # 214

9				6	7	5		
8	7		3	5			1	
						4		2
	1	8			3	2	9	4
7	9	2	1			3	5	
1		9						
	4			3	9		6	7
		7	6	4				5

Easy # 215

		9			8		6	
6			9	2		4		1
1			3	4			9	8
	3	4	2	5				
				3	4	9	5	
5	2			8	3			9
9		6		7	5			3
	1		6			7		

Easy # 216

	3		6	4		1	2	7
7	2		3					
4							6	8
5		1	4	9				
	8		1		3		9	
				5	8	4		1
8	9							6
					2		3	9
6	5	3		1	7		4	

Easy # 217

	5		9	1		6	3	8
		9				7	1	5
		1	3			9		
				9	6		5	
			4		8			
	9		7	2				
		4			1	8		
2	3	8				1		
9	1	7		8	3		2	

Easy # 218

	8		4		1	2		
			7					9
2		9		6			1	7
	6	2	8	7				
3		8				9		2
				3	5	8	7	
1	3			2		5		8
9					3			
		5	1		7		9	

Easy # 219

	3		2	6		7		5
7			3		1	8		
9							3	
2	8				9		5	
6		5		3		9		4
	1		7				8	3
	9							8
		1	6		8			9
8		7		4	3		2	

Easy # 220

4		1				2		7
			5			1	8	
	6		1					9
5		4	9				2	
6		2	4		5	9		3
	8				3	4		1
8					9		1	
	7	3			4			
1		5				7		8

Easy # 221

7		2		1			6	
					5			8
1		6			9	4	7	
			3	5		9		6
		5	9	8	6	2		
6		9		4	2			
	7	1	8			6		3
3			5					
	9			6		7		4

Easy # 222

1		8	9			3		
5		7	8	6	2			
	4		3	1		8		2
4						1	3	
	7	1						9
8		3		9	6		1	
			1	5	8	9		3
		4			7	5		8

Easy # 223

						2		
5	7		4	6			8	9
1				8	5			
3							7	1
7	2	1	8		6	3	9	4
8	4							2
			5	4				8
4	8			9	2		3	6
		5						

Easy # 224

							9	1
9	1			8	6			3
5	3				9	7		
			8	9		4		6
	9		3	7	4		2	
7		4		6	5			
		5	9				1	2
2			6	4			8	9
1	6							

Easy # 225

8			7	2		6		
3		9		8	6			5
				4	9			1
		1					2	
6	3		2	7	1		9	4
	9					1		
5			8	1				
1			4	9		8		6
		2		3	5			7

Easy # 226

7			2	6	8		3	
2	6						1	
	3	5			1	8		
	7			1		3		
1		9				6		2
		3		2			4	
		1	4			7	8	
	5						9	3
	9		1	5	3			4

Easy # 227

6		4	7		8		1	
8		5		6		7		
	7		4		2			8
3	4			9				
			1		3			
				2			7	9
7			2		9		6	
		2		8		9		7
	3		5		7	2		1

Easy # 228

		3	7					
5		9		2	8		4	
			4	5		8		9
				8		1	9	2
	8	6	5		9	4	7	
1	9	7		3				
6		8		7	5			
	7		9	6		3		1
					3	5		

Easy # 229

7					6		2	
2		1				5		7
	5	4			3			
	7				4	3		2
9		8	3		1	6		4
1		3	6				8	
			1			2	7	
3		2				8		5
	9		2					6

Easy # 230

		5			3			7
1		4			5		3	9
3	7	9		8	4			
	4							3
2	6	8				5	7	1
7							4	
			5	2		3	9	8
8	9		3			6		5
5			1			7		

Easy # 231

6			5				9	
	3			9		2	4	
	2		3	6				1
8		7	6					4
	5		7		9		3	
4					1	9		7
3				2	6		1	
	7	2		3			6	
	4				7			2

Easy # 232

	7		1	3	5		6	2
6	8		7				5	
		5			8		4	
	3	1		6	2			
	4						9	
			4	7		2	3	
	5		2			9		
	1				6		8	7
4	6		8	1	9		2	

Easy # 233

		7	4	1		5	9	
5					2			4
	8		5	9				6
		6			4		5	
	9	3		6		8	2	
	2		8			6		
3				4	8		6	
9			6					3
	5	8		7	3	4		

Easy # 234

		8			7			2
	3		5	8		7	6	
	6		3	1			5	9
5		6	4	3				
				5	9	4		3
6	1			4	3		2	
	2	4		9	6		7	
7			1			6		

Easy # 235

6	7						2	
9		3		7			5	
		5		8	4			3
3	1		9	6				
	5	6		2		8	3	
				4	3		6	2
8			2	1		5		
	6			5		3		4
	9						8	6

Easy # 236

		7			9	8	5	
				7	1			6
	4		6	5			9	2
	3		9		7		2	4
				2				
9	1		4		6		8	
4	9			1	3		7	
7			5	9				
	5	8	7			3		

Easy # 237

7		8						6
2	9		3	4	6	5		
			9		7			
5	1			7	2		3	9
	2						5	
8	6		5	3			4	2
			2		9			
		2	8	6	3		7	4
9						2		5

Easy # 238

		6	2	8				
2	7			6	9	5	8	
4								
	1		3				5	8
	2	8	4		1	7	9	
6	4				8		2	
								1
	6	4	5	1			7	9
				4	7	8		

Easy # 239

		3			7			6
4	8	9		2	6			
6		1			9		8	4
	5							3
7	3	6				4	1	2
9							5	
8	9		6			5		7
			5	4		8	3	9
3			9			6		

Easy # 240

	3			5	7	8		4
7			3		4		2	
	9	4		8				
		6	8			4	7	
	1			7			5	
	2	7			9	1		
				3		6	9	
	7		4		6			8
8		9	2	1			4	

Easy # 241

5	7						8	
		9	8					2
2	8		7			1		
8		7		6		5		1
	1		5		7		9	
3		5		2		4		8
		8			6		1	9
6					4	8		
	5						3	4

Easy # 242

	3		4			2		
7	2			6	3		8	1
	6			2	8			
				8	9	6		7
3								5
6		9	3	7				
			1	4			5	
8	4		2	9			6	3
		6			7		4	

Easy # 243

	1		8	6			9	
7	6						5	1
			5		3			
4	3		1		6		7	8
		5		3		1		
1	7		2		8		3	6
			6		2			
3	4						6	2
	9			7	1		4	

Easy # 244

7					5	2	6	
				1		7		4
	5	4		7	6		1	
2					7			
1		3	6	5	2	4		8
			9					5
	3		8	4		6	9	
4		7		6				
	6	9	7					1

Easy # 245

	7		2		5	8	3	
		8					2	
			9	8	1	7		
8		2	5			6		7
		9	6		3	4		
7		6			8	9		3
		3	4	5	6			
	8					5		
	5	4	8		7		9	

Easy # 246

5	9	8		2	3		4	
7	2	4						3
3					9			2
	4		5	3				
			8		6			
				1	7		3	
8			2					6
2						1	9	8
	1		9	8		3	2	7

Easy # 247

	2		5	9		3		
		6		2	3		8	
		5	4		7	9		
		9		3	1	6	4	
4								8
	6	2	8	5		1		
		3	2		9	8		
	9		6	1		4		
		7		4	8		6	

Easy # 248

		4	2	5	8	7	3	1
1	5		6				4	
				3				5
	7			4				
6		1	8		7	3		4
				6			2	
8				2				
	9				6		7	8
2	1	7	9	8	4	5		

Easy # 249

		4	7				3	
7		6	3			2	8	
			8	5		7	6	4
		7						8
4	3	2				9	5	1
8						4		
6	7	5		9	3			
	1	3			7	5		6
	4				2	3		

Easy # 250

9	6		2	4		5		8
						4		
				6	8		3	
6		9			3			1
1	3		6		4		8	5
4			7			3		2
	9		1	3				
		6						
8		1		9	5		2	3

Easy # 251

	9	8	7				4	
4					9		2	
	7		3	5	4	1	8	
3	5			8	1			
	2						6	
			2	7			5	1
	8	2	9	3	6		1	
	4		1					6
	3				8	7	9	

Easy # 252

			9	3	7	4	2	
		5			6	7	8	
	9	7		8	5			1
5		8					1	
	6					8		4
8			3	6		5	7	
	7	2	4			1		
	5	6	7	2	8			

Easy # 253

9			1	3	2	4		
		4						
6			4	7		5	9	
2	8		7					9
3		1	5		9	6		4
4					3		8	1
	3	9		4	5			7
						9		
		2	6	9	7			8

Easy # 254

	5		1	8		7	2	
	3		7	6			5	8
7					2	3		
8		1	6	4				
				1	8	4		7
		5	3					9
3	7			9	4		1	
	4	6		2	1		7	

Easy # 255

6			9			4	1	
		7		2	6		9	8
	8		3				5	
5	7					2		3
		6				7		
2		8					4	5
	2				4		7	
9	6		5	7		8		
	1	3			8			4

Easy # 256

	6	7		9				4
					2	3		
	4	9			1		5	7
			8	2		4	1	
	2		1	3	4		6	
	1	4		5	6			
7	9		3			8	4	
		8	2					
1				4		5	7	

Easy # 257

	7				2		1	
5	4	1					7	
3	8	2		1	7	5		
		5	8	7				
			3		6			
				9	4	7		
		9	2	3		1	4	7
	1					2	3	9
	3		1				6	

Easy # 258

1		7	5				4	
6		5				7		
	9		7					1
7	5			3			6	4
		4	6		5	9		
2	6			1			8	7
3					8		7	
		6				2		8
	7				3	4		9

Easy # 259

9	6		7			8		4
				6	8	5	7	
8	5		1				6	
	7	2	8			6		
	8						9	
		6			2	1	3	
	1				6		5	7
	2	5	3	1				
6		9			7		1	3

Easy # 260

3	1		5					
			3	8		5	2	
				7	4		3	
6	3		8				7	2
		8		6		1		
9	4				7		6	3
	9		6	4				
	7	4		5	8			
					2		5	4

Easy # 261

1			5		6			
				2			3	
7			1		3	6	4	
9	1		2			5		
	2	4		5		3	9	
		3			4		2	8
	8	1	4		5			9
	9			7				
			8		1			3

Easy # 262

	7	4		3		5		
			9	5			6	
			4		7	3	8	1
	4	7		2	9	1	5	
	5	3	1	8		9	7	
4	1	2	6		8			
	9			1	5			
		5		4		6	1	

Easy # 263

1		5	4				2	
		3	8	6		1		5
4	6			5				
			1				7	
3	2		5	7	9		8	6
	9				4			
				2			6	4
6		7		4	5	2		
	4				7	5		9

Easy # 264

	4	8	7		9	5		
	5		6	4	2			
		9					4	
5	1				7		9	4
	3		8		1		2	
8	2		4				1	5
	7					4		
			1	7	3		8	
		2	5		4	7	3	

Easy # 265

		2		1	6	8		
8		7				1		5
			3		7			
6		5	1		8	3		4
	8			3			7	
1		3	6		9	5		8
			9		1			
9		1				4		3
		4	8	5		2		

Easy # 266

		1						
	4	6	2		5			1
2		8	1		9			4
		5	9			6	2	
8	2		3		6		1	7
	9	7			1	4		
7			5		8	3		2
5			7		2	1	9	
						7		

Easy # 267

8	6				2	7		
							8	3
5	2			9	3		6	
			7	3		9	1	
6			9	1	4			2
	3	9		2	5			
	4		3	5			2	8
2	8							
		1	2				7	4

Easy # 268

5					8	1	2	
		4	9	6				
	2		1	5	7		4	
3	7		8	4				
2	4						1	8
				1	3		5	4
	6		5	9	4		3	
				2	6	5		
	5	2	3					7

Easy # 269

	3							4
		6	3	4		2		8
5			2	1				
	9	2	6			5		1
	8	7	4		2	3	9	
6		3			1	4	7	
				8	6			3
3		9		2	5	8		
8							2	

Easy # 270

	9			1	5	4		
		5	3	7				
		7	2	6		8	3	
6							7	
1		8	9	4	7	2		6
	7							9
	6	1		3	8	5		
				2	6	7		
		3	4	9			8	

Easy # 271

	1	4				3	8	
9	8				2			
		5			8			6
1					5	2	3	
	5	7	2		3	6	1	
	3	8	7					9
8			5			9		
			3				7	4
	4	9				8	2	

Easy # 272

		3	7	5		1	8	
	8		2	3				
5							3	
4		6	9			8	2	
1		8	5		6	4		3
	9	7			2	5		1
	6							8
				9	5		7	
	3	5		6	8	2		

Easy # 273

			7				2	
7	9		2			6		3
	1				6	7		9
		1				9	5	4
	2	5				3	6	
6	7	9				2		
2		3	8				9	
9		4			2		7	8
	8				4			

Easy # 274

			6		8			
7			5	4				2
8	6						1	7
1	8		9		6		5	4
		5		1		3		
4	9		8		5		7	1
3	5						4	8
2				8	9			5
			1		3			

Easy # 275

			5	4		7	8	9
8		9	3				5	2
7			9				3	
9						5		
2	3	7				6	4	1
		5						7
	7				2			3
3	6				9	8		4
4	9	8		1	3			

Easy # 276

1		3		5	4	9		
	4	8		1	7	3		
	9		8					4
				6	5		7	1
4	6		1	7				
2					9		3	
		4	7	8		6	5	
		7	6	2		4		9

Easy # 277

4	9		7		2			
	2	5		8	6	9		7
7						8		4
	3				8	5		6
	5						7	
6		7	2				8	
3		6						1
5		9	3	4		7	6	
			5		7		9	8

Easy # 278

				1	7	5		
	4	7		2	5			1
		9			3			2
8	5			9		7		
9		6	1		4	2		5
		1		7			4	9
7			6			3		
1			3	8		4	9	
		8	7	5				

Easy # 279

	2	9	4	1			6	8
	8	4				7		
			9		6	5		2
	6	8	3					5
9								6
4					5	8	9	
2		1	6		3			
		6				1	5	
3	9			5	8	6	2	

Easy # 280

8				9		7		1
6			8	7			9	
	7		1					4
	1	5	6				4	
9			5		1			2
	4				8	1	3	
5					2		8	
	6			8	9			7
4		7		5				9

Easy # 281

	2		7	5			8	
7		3			1	9	2	
				4		7		1
2	3							7
5			3		4			6
4							1	3
8		4		3				
	7	5	8			6		9
	6			9	7		3	

Easy # 282

9			7	2				
					6	2		4
8	2			4	3			
2		9			8	5		7
	3			7			1	
5		7	3			6		8
			5	3			4	6
1		5	4					
				8	2			5

Easy # 283

5	3			1	4			8
				3		1	7	
		6			7		5	3
		2			5			
	1	4	9	2	3	6	8	
			7			9		
3	9		2			7		
	7	1		6				
6			3	7			1	2

Easy # 284

3	4		5			1		2
	9			7	1		4	
1		5		6				
		1				4	2	
		8	6		2	7		
	5	2				6		
				2		9		6
	2		1	3			8	
8		3			9		1	7

Easy # 285

	3		6		9		8	
				3				
	2	1	5		7	4	9	
7			9		6	3		8
		9		7		5		
2		6	3		8			9
	9	3	7		5	1	4	
				2				
	5		4		3		7	

Easy # 286

9		2				4		
			6		8	7	3	
3		6	2	5			9	8
8		9	1				7	
	6						8	
	2				7	9		6
6	1			7	9	8		3
	3	5	8		1			
		8				5		7

Easy # 287

		9	8				2	6
	1		9				3	
8				1	4	7	5	
	9	4				3		8
1								5
3		6				4	1	
	7	8	5	3				1
	4				6		8	
9	2				7	5		

Easy # 288

		8		3	6	1		
1		4	9	2	7	3		
				8			9	
		9		4				5
2		3				4		1
5				1		2		
	5			7				
		2	1	6	8	5		7
		1	2	9		6		

Easy # 289

	7	6	9			1		5
	2	9	1				7	
				7	6		4	2
7	6		4					1
	5						3	
1					3		9	4
2	9		3	1				
	1				7	3	2	
3		8			9	5	1	

Easy # 290

				5	8		1	
						4		
6	5		2	4		3		8
5		6			1			9
9	1		5		4		8	3
4			7			1		2
8		9		6	3		2	1
		5						
	6		9	1				

Easy # 291

				3		1	2	
4	1		6	8		9		
		5	9		2			4
	5	6			1		8	
		8		5		7		
	2		4			5	9	
5			3		9	6		
		3		7	5		4	9
	9	1		4				

Easy # 292

			6	8	1		2	4
7	8			3		6		
6					2			
	9			6	5		4	7
4	5						6	8
8	1		4	7			3	
			2					5
		8		1			9	6
3	6		9	5	8			

Easy # 293

5		1		7				
9	2		1			3		6
	6			3	2		7	
		5				7	4	
		9	7		5	6		
	7	8				2		
	8		2	9			1	
7		2			4		8	3
				5		4		2

Easy # 294

				3	9	4		
		8			2			9
	5	1		4	8			7
5	1			9		7		
3		6	1		7	2		5
		9		5			3	4
7			3	6		9	1	
6			8			5		
		3	9	7				

Easy # 295

4			7		3	2		
9		3		8	5		6	
	2					3		
	4		9			5	3	
1		7		5		8		2
	3	6			2		1	
		2					5	
	5		6	7		1		9
		9	5		4			3

Easy # 296

4			7		2	8		
		9	5		4		7	1
	7			3			2	4
				7		4		2
			1		9			
9		6		2				
3	5			8			4	
8	6		4		3	1		
		4	6		7			3

Easy # 297

			4		6			2
	3	4	1		2			7
	1			5				
		6			5		2	8
	8	1		6		3	5	
9	5		3			1		
				7			8	
8			6		3	2	9	
1			2		9			

Easy # 298

				1				
	6	8	9		7	5	3	
		4	8		2	1		
4	1		2		8			9
	7			9			8	
8			4		1		2	5
		9	1		6	7		
	3	6	7		9	8	1	
				5				

Easy # 299

7	6		5		4	1		
		5	6		3			4
4	9			7			5	
8		6		2				
			1		8			
				3		5		2
	3			4			2	5
5			3		2	7		
		8	9		5		3	1

Easy # 300

1	8	2			7			
6			4					7
	4	9			1	3		
8	1			9				5
	7			4			2	
9				5			7	1
		8	2			1	4	
4					6			2
			9			6	3	8

Easy # 301

	5			4			7	
				7	6	1		2
7	2	1	3				6	4
		2		5	4			8
		8				2		
3			2	1		7		
2	6				9	5	1	7
8		7	5	6				
	1			2			8	

Easy # 302

9			6	1				
		1	2	5		4		9
	5							1
	8	3	7			9		6
	4	9	5		3	8	1	
7		2			6	5	4	
3							9	
1		5		3	9	6		
				7	5			2

Easy # 303

	5		4					
7		4	1	5		8	9	6
2						4		
8		5			4	6	7	
		3	2		8	1		
	4	7	9			3		2
		9						8
3	6	8		9	1	2		7
					3		6	

Easy # 304

3				4	6		1	
5			7					6
	4	6		8	5	2		
	6		5			3		
	7	1		3		9	4	
		3			1		7	
		5	9	2		1	6	
9					3			4
	3		1	5				9

Easy # 305

5			6	1		3		
7		2		4	8			9
				9	2			6
		9					4	
8	4		9	5	3		1	7
	3					9		
9			4	8				
6			7	2		4		1
		7		3	5			2

Easy # 306

		1	4	9		5	2	
	4				8			1
		5	3	2		8		4
3	2		9	6				
				3	2		4	6
9		6		8	3	4		
5			1				7	
	1	4		7	6	3		

Easy # 307

	8		9		1	6	5	
			2	6	7	8		
		6					9	
6		9	1			3		8
		2	3		5	4		
8		3			6	2		5
	6					1		
		5	4	1	3			
	1	4	6		8		2	

Easy # 308

3		9				7		1
7					9		5	2
6			7	5			3	
8					2		1	
	1	5	3		6	2	7	
	6		5					4
	7			9	5			3
5	9		4					7
1		3				6		5

Easy # 309

7	1	5		9	8			6
					4		2	5
2	3						1	
				7	3	6	9	
3			9		5			2
	7	9	6	2				
	6						3	1
4	8		5					
5			1	6		9	8	4

Easy # 310

1				7	3		9	
6		8	9	2			5	4
				6		8	1	3
5	8	4						
	2						6	
						5	3	8
7	5	9		8				
8	6			4	5	9		1
	1		3	9				5

Easy # 311

9					2	6		
	3		8	1		4	5	
	8		5	9			3	2
3		5	7	8				
				5	4	8		7
7	6			4	3		2	
	1	3		7	8		6	
		2	1					3

Easy # 312

1		4	9				3	
				6	5	8		
8			2	3	4			1
				8	9		7	2
4	9						1	8
3	8		7	4				
7			8	5	3			6
		3	6	1				
	2				7	1		3

Easy # 313

7	9			2	6		8	
				8	7		3	
	4		3	1				5
8						2		
	6	2	8	4	5	1	9	
		5						8
9				5	4		7	
	8		2	6				
	3		9	7			1	2

Easy # 314

				5	7	2		
	8		2	9		4	3	
5					3		9	6
	1		3		5	8	4	
				4				
	7	3	8		2		6	
6	9		5					1
	3	8		7	1		5	
		5	9	3				

Easy # 315

1				6	3			8
				1			7	
4		8	7	2	5			6
7				4		9		
6		2				8		4
		9		8				2
2			8	3	1	5		9
	9			5				
8			2	7				3

Easy # 316

1		8				2		7
			2		3			
		7	5	8		4		
9		3	7		8	1		5
	2			3			7	
7		1	6		5	3		8
		4		1	7	9		
			8		6			
3		9				8		6

Easy # 317

6			2	5	4	9		7
9			1	6				3
	4			3				
		8		7				4
7		9				5		6
5				9		8		
				2			8	
1				4	5			9
8		2	3	1	9			5

Easy # 318

		4	6	7		9		
6	3				1	4	2	
				8			6	1
4		3						6
7			3		8			5
8						1		3
9	8			3				
	7	6	9				5	2
		5		2	6	3		

Easy # 319

		2				3		5
	8	3	2		1			
7	1			5	4	2		8
	9				5	4		7
	7						2	
2		4	1				5	
8		7	9	3			4	2
			7		2	5	8	
4		9				6		

Easy # 320

	8			3	9	1		7
	4	7		1				
9			8		7		2	
		5	1			7	9	
	6			9			3	
	2	9			4	6		
	9		7		5			1
				8		5	4	
1		4	2	6			7	

Easy # 321

2			7	8	6			5
7		3	5				9	
				3	2	7		
				1	5		6	7
6	3						4	1
9	5		4	6				
		6	8	2				
	7				4	1		3
3			1	7	9			6

Easy # 322

2		3		8			9	5
				2	1			
1			5			7		
		6		7			3	4
3	4	7		6		2	5	1
8	1			5		9		
		8			5			3
			7	1				
9	5			4		6		8

Easy # 323

5	6					2	4	8
						6	7	
		3			6	5	9	
9		6		8	2			
		5	6		7	4		
			1	4		9		5
	5	9	3			7		
	3	4						
6	2	1					5	4

Easy # 324

6	2	8	3	9	4			5
	9			8				
		5			7	9	2	
				5		6		
8	5		6		3		7	2
		4		7				
	3	6	7			1		
				4			3	
9			5	3	1	2	4	6

Easy # 325

		1	9	4			2	7
2	9			7	1			
					5		6	
4	3	2		9				
	1	5	2		7	9	8	
				6		2	5	3
	7		6					
			7	5			9	8
3	6			8	2	5		

Easy # 326

		1	5					
			2	4		6	7	
	4	7		8	6			2
				6		3	8	7
6		9	4		7	2		5
7	3	5		1				
5			7	9		1	3	
	9	6		5	4			
					1	4		

Easy # 327

8		2	7		6			
		5	8				6	2
	9		2		5		8	
1			5				4	
9	4						3	1
	2				4			8
	1		3		2		5	
3	8				9	7		
			4		7	9		3

Easy # 328

2			6		3	5	4	
	3			5		6	8	
		6	7		4			3
				1		9		4
			9		2			
3		1		7				
5			1		7	3		
	1	3		6			7	
	7	2	3		8			9

Easy # 329

8	9			4		3		1
			2	5				
		1			9			6
1	5			9		8		
6	4	2		3		7	9	5
		3		2			6	4
5			9			2		
				7	5			
7		6		1			8	9

Easy # 330

	8		5			2		
5	4		2			1	9	
			1	6		4	5	8
	5							1
8	9	2				6	7	3
1							8	
4	6	5		7	2			
	2	3			5		6	4
		8			9		2	

Easy # 331

9			3	1				
		1	4	2		8		9
	2							1
	6	5	7			9		3
	8	9	2		5	6	1	
7		4			3	2	8	
5							9	
1		2		5	9	3		
				7	2			4

Easy # 332

7						4		6
9			3	6	4	8		
	3		7				1	9
	9			7				8
	6	4				7	5	
2				4			9	
3	8				2		7	
		2	9	1	7			5
5		9						1

Easy # 333

3		4				1		
6	5		4		7		2	
1		9	2	5				
	3					4		2
			7	9	8			
8		5					9	
				2	4	9		6
	9		8		6		3	4
		6				8		5

Easy # 334

5	2			3	7			
		3	8	2			9	6
					6			7
3	9	8		6				
2		5	7		8	3		4
				5		8	1	9
6			3					
8	7			1	5	4		
			4	7			8	5

Easy # 335

	6		5	2		9		
		5	1		4	3		
		2		8	6		1	
		8		7	5	4	6	
5								2
	2	6	8	3		1		
	5		3	4		6		
		1	9		2	7		
		3		1	7		4	

Easy # 336

	6	4	8	2			7	3
				4	7	1		
								2
6	4				1		9	
	9	1	4		2	7	3	
	2		5				8	1
4								
		6	9	1				
9	7			6	3	8	1	

MEDIUM

Puzzles

Medium # 337

	2						5	
4				2	5			6
		1		3				
			7	6		5		
	8	2				7	9	
		4		1	2			
				5		4		
9			1	8				7
	3						1	

Medium # 338

9			5			1		
			7					
		4	6				5	9
2		7					6	
4	3						8	7
	1					9		5
8	7				9	3		
					7			
		2			5			4

Medium # 339

								9
	2	7					3	
1			6	2			5	
					7	6		4
		9	5		6	7		
7		4	9					
	7			9	1			5
	8					2	6	
5								

Medium # 340

		1	2					
	5			7				2
				4		7	3	
	4	7			9	2		
	6						9	
		9	6			5	4	
	3	2		6				
8				1			6	
					3	9		

Medium # 341

7				1				
	8	2				3		
			7			4		
1	5		3	6			4	
	2						5	
	6			2	4		7	9
		3			8			
		1				9	3	
				3				6

Medium # 342

4		8			3			
	9			5				
		1				3	5	
6	4		8			7		
7								4
		3			4		1	5
	3	7				5		
				7			2	
			3			6		9

Medium # 343

			6	8			4	2
		6				9		
	2		4		1			
4		1						9
				6				
7						4		8
			5		6		2	
		2				3		
1	5			9	7			

Medium # 344

			4		1			
	7					9	3	
	5	2			9			
8							5	6
		5	1		7	4		
1	2							3
			9			2	7	
	4	3					6	
			6		4			

Medium # 345

	6		8	4		5		
7								
	5	8	9					7
6				5			9	
			2		1			
	2			3				4
2					8	3	4	
								9
		6		2	5		1	

Medium # 346

						9	5	6
		5	6				3	4
7			9					
	3				4			2
			3		8			
8			5				4	
					3			8
9	2				6	5		
6	7	8						

Medium # 347

			5		1		9	
	7		9					8
				2		4	3	
				9		8		1
			4		8			
2		8		6				
	3	4		1				
9					3		1	
	5		6		4			

Medium # 348

					1	5	4	
2	1							
7				9			1	
5		6		7	9			
	7						9	
			5	3		6		1
	2			4				6
							5	3
	4	3	9					

Medium # 349

2	9	3						
6				3				
		7	2		9			
		9		8	1			
4		1				2		7
			6	7		1		
			8		3	7		
				6				2
						6	3	1

Medium # 350

	5						8	
	4			9	2			7
						1		
4		7		3	9			
8								6
			7	1		8		3
		6						
7			3	4			5	
	2						1	

Medium # 351

	7	5					2	
				7				5
2				6				
	6		3					4
8		2	9		4	7		1
7					8		5	
				4				6
9				3				
	4					3	8	

Medium # 352

2		4						
5	6			8	3			
7								
		8	1		7			6
	2		6		8		5	
3			4		2	9		
								1
			5	2			4	7
						6		3

Medium # 353

				4		3	5	8
		8	5	3			9	
					2			
1	3			8				
7								3
				9			6	1
			3					
	2			6	1	4		
6	9	7		2				

Medium # 354

		7						5
	3				4	9		
5		4	1			3		
4			2					9
	6						8	
2					1			3
		5			2	4		8
		3	8				9	
6						7		

Medium # 355

		3			2	8		
2	4		5					
6		1				5		
	5			6				
	2		8		1		3	
				7			1	
		4				7		8
					6		5	4
		5	3			6		

Medium # 356

	5						3	7
8		9	3					
7			6		8	4		
4		2						
	9						7	
						6		2
		8	9		1			4
					6	5		1
2	7						9	

Medium # 357

						7		3
2								
	9	3		5		4	2	
	1	7	6					
	5			8			9	
					5	8	3	
	8	4		7		3	5	
								2
7		9						

Medium # 358

2						5		6
		9			4			
	5		9					
6				8	1			4
	7		5		9		6	
5			4	6				2
					3		4	
			1			6		
1		4						3

Medium # 359

	6	1			3		8	
					6			1
	3					9		7
9				2				8
			6		9			
3				4				5
7		3					1	
2			7					
	5		1			7	3	

Medium # 360

				1				2
			7				8	3
	9				8		5	
9		1	8			7		
			2		6			
		6			9	3		8
	2		9				3	
5	8				1			
1				3				

Medium # 361

2		1						
			7	6			5	
			1			3	9	4
	9	3				1		
7								9
		8				5	4	
3	2	7			4			
	6			9	7			
						8		7

Medium # 362

		5	7					8
			4		9	7		
	3	7		6				9
7	6							
			8		2			
							9	7
3				4		2	5	
		2	5		8			
5					1	6		

Medium # 363

	3		1				5	
					3			
	8	7	2	5				1
	4			3				9
		2				3		
1				8			7	
3				7	6	8	1	
			9					
	1				4		6	

Medium # 364

3	1		4			2		
	8		5	9				1
							6	
	2			1				5
9				6			2	
	7							
4				3	2		5	
		2			4		1	8

Medium # 365

					5		7	
		2						9
	6		2	1				5
	9			6	1	8		
		6				1		
		1	7	9			4	
3				8	7		1	
7						5		
	1		9					

Medium # 366

7				9	8	2		
	9							7
		6				5	4	
				4				3
5		9				7		2
8				6				
	3	4				6		
6							2	
		5	2	3				8

Medium # 367

		4		9	2			
				1				
3	7					8		
4	6	1					2	
		2		7		5		
	9					1	6	3
		5					4	6
				8				
			9	4		3		

Medium # 368

8			2	4		6		
5			9				7	1
						4		
		9			4		8	
2								6
	3		5			2		
		8						
6	9				5			8
		2		6	9			4

Medium # 369

		1		2			7	
	8	4			3			
9						1		8
2				1	8	4		
		9	4	7				2
5		2						4
			3			7	5	
	3			4		9		

Medium # 370

		4			2	1		
		9					5	
1			4		6		7	
			9			7		1
		7				4		
2		1			5			
	7		3		1			2
	2					8		
		5	7			3		

Medium # 371

			4		2		5	
						8		
	3	9	7	8				
		3			6	5		
	9	8				6	2	
		1	3			9		
				9	5	7	6	
		5						
	1		2		7			

Medium # 372

1				2			3	
		5					6	
6						4		7
			5				7	6
		3	4		9	2		
8	6				1			
9		7						3
	8					5		
	5			8				9

Medium # 373

		6		2	3	4		5
	8		5					
			6				7	
			1					6
		4				9		
3					8			
	2				4			
					5		2	
9		3	8	7		6		

Medium # 374

			8	6				
		2					8	
			2		4		3	6
8					6		4	9
9								3
6	1		7					8
1	7		9		8			
	5					3		
				4	3			

Medium # 375

4	5	9	1				6	
							4	
8		7	5					
				5		4		6
			9		2			
2		6		8				
					4	8		2
	3							
	8				7	1	5	4

Medium # 376

	9		1		8	2		
7				4				
	2	3					9	
	1	9						3
			3		2			
4						1	5	
	7					6	1	
				5				7
		8	7		3		2	

Medium # 377

	3							
	5		8		9			
	1				7	9	4	
5		7		9				4
				6				
2				1		7		5
	9	8	5				6	
			6		1		3	
							7	

Medium # 378

5	2				3			
3	4			9				1
		1			5			
4		8		5				
	9						6	
				7		5		3
			4			2		
9				2			4	5
			5				7	8

Medium # 379

			9		1			6
	9			3				
	5	8			7		9	
7		9		8				
		4				5		
				2		3		7
	7		3			1	4	
				9			5	
4			8		2			

Medium # 380

		7	5		3			9
	2		1					5
		1		6			3	
		6			4			
1								6
			8			7		
	7			4		5		
6					2		4	
2			7		5	1		

Medium # 381

1	8			6	5			
	2	4					6	7
3				7				
	9	3	1					
					8	2	9	
				2				6
7	4					1	8	
			9	1			7	5

Medium # 382

		8			7		1	
5			9			8		
2					6			7
8				7				
	7	4				3	5	
				3				2
1			2					3
		5			4			9
	9		8			5		

Medium # 383

					6	1		2
5			4					
	7	1				5		
1					4	2		
	3						6	
		9	8					1
		4				9	7	
					1			6
8		3	9					

Medium # 384

		1					9	
	8		5	2	3			
		2						
	4	5	3				1	8
			9		1			
8	1				2	3	7	
						1		
			8	6	7		4	
	9					7		

Medium # 385

4			2	9				
3		2		6				
1			4				8	
				8				4
	8		1		9		2	
7				3				
	4				7			1
				1		5		8
				4	5			9

Medium # 386

			6	5		8	2	
5			8			6		7
							4	
6						7		
3			4		6			8
		9						4
	4							
1		6			3			5
	5	2		8	9			

Medium # 387

						2		3
		9					1	
	8		4	6	3			
		4	6			1	9	
2								5
	9	7			5	4		
			1	4	7		2	
	2					6		
7		5						

Medium # 388

		4				2		1
	3			2			5	4
			7				8	
			3	8	5			
2								6
			2	4	6			
	9				7			
3	2			9			7	
4		1				8		

Medium # 389

1			7		4			
7								
	9		3	1			4	
8			4			5	7	1
4	5	3			1			6
	7			6	2		3	
								9
			5		8			4

Medium # 390

	9	2		4				
							3	4
	1	7		3		2		
	4		9					
	8		2		4		7	
					3		5	
		9		2		8	1	
1	6							
				7		5	4	

Medium # 391

	1	3			2		5	
9	2			5				
		7			8			
	3				4	7		
		4				8		
		1	3				6	
			9			2		
				8			4	5
	4		7			9	3	

Medium # 392

						2		
3			4		9	8		
					8		3	4
				4			2	1
		5	6		2	3		
1	8			5				
5	6		2					
		8	1		5			7
		4						

Medium # 393

	7						6	
		3					5	1
		2	8	4				3
			2	8				
9			5		3			6
				6	1			
1				9	6	7		
8	6					3		
	2						9	

Medium # 394

				5			9	6
	8	3						
			2	4	8			3
9	4		8					
		8				5		
					9		7	8
1			7	9	6			
						1	6	
5	3			8				

Medium # 395

4				2	1	6	9	
				4	6			
			3				4	
			9			3		
3	8						7	6
		4			8			
	1				5			
			2	1				
	5	3	8	7				9

Medium # 396

1	2				9			
5			2					6
		9			7		8	1
				9		7		
	5						4	
		3		4				
8	1		7			6		
6					5			4
			6				5	3

Medium # 397

				8		9	1	
		1	2					
	9		5					2
	5				7	8	4	
4								7
	2	3	6				5	
3					5		8	
					2	6		
	6	5		1				

Medium # 398

5	8	1	9				6	
		9		7	3		5	
9		6					4	
	5						9	
	2					1		7
	9		7	6		2		
	4				2	8	7	5

Medium # 399

4	7	9		1				2
2			7				3	
				2			5	
		1			9			
			4		5			
			6			3		
	3			4				
	1				8			3
7				6		5	9	1

Medium # 400

					7	6		
		6		8				3
			6	3				4
		3		2	6		5	8
9	4		8	1		3		
4				6	1			
7				4		5		
		8	9					

Medium # 401

			5			6		2
5			3	2	6			
		2	8			4		
						2		7
1								3
7		9						
		1			5	9		
			2	9	1			6
6		7			8			

Medium # 402

8	9							3
		1			6			
		7		4			2	
	7					5	6	1
3								8
1	6	5					3	
	5			3		2		
			9			6		
6							8	5

Medium # 403

	1			6	9			2
	2		5					
		6				3		4
9			1		3			
		5				6		
			7		6			5
7		9				2		
					5		8	
8			3	7			4	

Medium # 404

			7	3		8		
						3	4	6
	4			9				
			6					5
5		7	3		8	9		2
2					7			
				7			1	
7	9	1						
		8		5	1			

Medium # 405

			5		6			
	8			3		9		2
6				9			8	
			3	2			1	5
9	7			1	4			
	3			5				1
8		9		7			3	
			2		3			

Medium # 406

				2	8		1	5
7				4				
			9			4	8	
						3	2	
2				3				7
	1	7						
	8	4			6			
				1				4
9	2		3	5				

Medium # 407

	8	7	6	4				
							2	4
		1			3	6		
	3		2				8	
			3		6			
	9				8		5	
		3	5			8		
4	5							
				8	1	5	3	

Medium # 408

		8		6				5
				2	4			
6	3		7				9	
9	6							
	7			4			5	
							2	9
	4				8		7	3
			4	9				
7				1		6		

Medium # 409

	2		7					6
					1	5		
9			5	2			4	
		6	8			3		
	4						5	
		8			3	2		
	5			1	7			8
		2	6					
3					4		7	

Medium # 410

			6		1			
4				3				5
	2		5			9		
3		4		6			1	
	1						2	
	6			9		5		3
		7			5		4	
5				1				7
			7		6			

Medium # 411

	2			3	9	6		
			1	2		9		
								3
3		9			2			
2			5		1			7
			8			2		5
1								
		5		1	4			
		4	2	6			7	

Medium # 412

4			7			5	1	
	1		4			6		8
				8				
		1	6				8	
		5				9		
	9				3	4		
				2				
6		3			8		7	
	8	9			5			6

Medium # 413

5			3				1	4
	9	3			6			
4					7			
3							4	2
		5				1		
6	1							8
			2					3
			6			2	9	
9	7				4			6

Medium # 414

							8	7
		2			6	4		
8					2		5	
			1			7	2	9
3	6	9			5			
	3		4					6
		4	2			9		
9	1							

Medium # 415

3	8	2		6				
	7					8		
		6	2		5	7		
						5	8	7
1	3	8						
		9	8		1	2		
		5					1	
				5		9	4	8

Medium # 416

7					6	5		
8								
4	2		3				6	
				9	3	8		
2				1				3
		9	7	5				
	7				1		9	8
								6
		3	5					2

Medium # 417

	3		5	2				
		6	3					
7	1	2						9
	2		9			6		
	6						8	
		7			1		5	
6						3	1	8
					5	4		
				4	8		6	

Medium # 418

			2	5		7		1
				3			9	
		4				8	3	
			4			3		
8		5				6		7
		7			5			
	1	6				5		
	7			2				
9		2		6	3			

Medium # 419

7			3			1	2	
	1							
			7		6	9		8
5	7							
			1	4	5			
							5	6
8		3	5		4			
							8	
	4	1			9			3

Medium # 420

								3
2	1				5			4
			4	2	6		5	
9	8					6		
		4				8		
		3					7	9
	2		7	8	3			
6			5				4	7
7								

Medium # 421

	9		2			6		
1				3	6			2
	6			5			4	
6						7		
7								4
		8						6
	5			6			8	
2			9	7				3
		9			5		6	

Medium # 422

8		2						
			4	5			1	
4			9		8	7		
				6		3	5	
		7				8		
	8	6		4				
		1	7		9			2
	3			2	1			
						1		9

Medium # 423

	7	8				5		
		1						3
	3		8	4	6			
			5					
6	9		7		2		1	8
					9			
			9	1	5		2	
8						3		
		7				1	4	

Medium # 424

			4			3		
				2		1		
1		8		5		7		
		7		4				9
	9			6			1	
3				8		5		
		2		1		8		6
		9		7				
		5			8			

Medium # 425

	5	4		9				
							6	
6	9		2	8				
7		6						
4		2		3		7		5
						6		1
				2	1		4	8
	4							
				6		3	2	

Medium # 426

		9	2				8	
6					4			
3				1	6			7
				2	3	9	6	
	8	5	6	4				
1			5	3				6
			7					8
	6				1	3		

Medium # 427

	6		3	1		5		
			9	4			2	
		8						
8					5	3	6	2
2	7	3	6					1
						1		
	9			6	7			
		7		5	3		4	

Medium # 428

8		6	3					
	4	5	9		7			
		9				5		
			2					7
3		8				4		6
2					3			
		1				6		
			1		5	7	4	
					9	1		5

Medium # 429

	1						2	
	4			1				
6			9				1	7
7			2			3		9
	5						6	
1		2			4			8
5	3				9			2
				4			3	
	8						5	

Medium # 430

7			5			4		9
3							1	
				9	4		2	
8		9		5				4
2				6		9		7
	8		3	2				
	7							8
5		1			7			2

Medium # 431

		1				7		5
		6	2	1				
	7				4			
7				2			1	4
		2				5		
9	1			3				7
			7				9	
				9	3	8		
8		3				6		

Medium # 432

9			5					8
		3		8				4
	1	4						
2	4				5	1		6
5		8	1				7	3
						9	5	
3				9		6		
6					8			7

Medium # 433

			2				9	5
		9		3				
3				4			1	
2				1			6	
		3	6		9	8		
	8			5				2
	6			8				3
				9		4		
5	9				6			

Medium # 434

7	1				9			2
	5	9			1	7		
2							4	
		1		3				
			9		4			
				6		3		
	4							8
		7	5			1	2	
8			6				5	3

Medium # 435

		9			2		5	
	7	4						8
					6		4	
				5			6	
8	5		6		7		2	3
	3			9				
	9		5					
3						5	8	
	6		1			3		

Medium # 436

				2		6	8	
		9	5		6			1
3	1			9				
				3			6	
6								5
	4			7				
				8			5	2
2			4		9	3		
	9	1		5				

Medium # 437

	5					6	8	
8				6	3			7
				1	5			
7			5			1		
2								8
		4			7			3
			2	9				
5			7	3				1
	7	3					6	

Medium # 438

7			4				1	
				8	2		7	9
		4			9			
					7		5	
		3	5		1	9		
	4		9					
			7			1		
6	7		3	2				
	5				4			2

Medium # 439

							4	
	2			8		5		6
			1					2
	4	6	5		9		3	
	9		4		2	8	5	
5					6			
3		2		1			9	
	8							

Medium # 440

6			5				7	
		2						
4		1	3	2				
3				9		8		
	4		1		6		9	
		6		3				4
				4	5	6		8
						5		
	6				1			3

Medium # 441

		7		9			8	
					6	7		
5		1					6	4
1	8		6		3			
			5		9		7	1
8	5					1		3
		6	2					
	1			4		9		

Medium # 442

			2				3	
	8				1			
			9		3		7	1
4				8				5
		5	6		9	7		
6				1				9
5	6		1		4			
			3				1	
	3				6			

Medium # 443

					6	9		1
6		4						8
	2					6		
			9	5			2	
	5		8		4		6	
	4			1	3			
		2					3	
5						7		6
1		8	7					

Medium # 444

		8					7	
				7	4		9	
	3	9						2
	1			2			3	
2			8		3			5
	4			6			2	
5						9	6	
	8		6	4				
	2					7		

Medium # 445

7		4						
			8				2	
	3			9		5		6
			5			1		4
	4		1		6		3	
8		3			2			
3		6		8			7	
	7				1			
						3		5

Medium # 446

					1			
	4	7		2		1		6
6			9			4		
			5			9	1	
	3	4			8			
		1			6			5
2		5		4		8	3	
			8					

Medium # 447

		7	9			4		
8	2		5				7	
					2			
	8	1						2
5			4		9			3
2						7	4	
			3					
	4				7		5	6
		8			5	9		

Medium # 448

			2		1		6	
					5	8		
2						4	9	
	2	6	8					
8				1				7
					2	3	5	
	3	8						5
		2	3					
	6		5		4			

Medium # 449

2							4	
				6	7	3		9
		6						8
8			3	7				4
	3						1	
1				8	9			5
7						4		
9		1	6	5				
	5							3

Medium # 450

	8	7		1	9			
						9		7
6				8	5			
4					6		5	
8								6
	2		5					4
			6	9				5
9		4						
			8	3		1	7	

Medium # 451

	2		6					
5					7			
			1			6		3
	5				1		9	4
4	3						2	8
9	1		8				3	
1		8			5			
			7					9
					8		5	

Medium # 452

2	5		1	4				9
		6		3		4		
5					1			7
	6	8				2	9	
9			7					3
		3		2		9		
1				7	5		8	6

Medium # 453

	7	6	9			8		
				7		1		
3				8				
	9		8		7	6		
	6						4	
		2	5		3		8	
				9				3
		4		1				
		7			5	4	2	

Medium # 454

	9		7			1		
	6		5		9			
		3		1		6	9	
					4			
7		1				5		3
			8					
	8	9		3		4		
			6		5		2	
		4			7		1	

Medium # 455

	1				2			7
7			3		9			1
	8	2						
	3			1			2	
				8				
	2			6			9	
						3	5	
5			1		6			9
9			4				7	

Medium # 456

	4					5	6	
			2				3	9
		6		4				2
	9		1		3			
		1				9		
			7		8		5	
3				7		1		
2	6				5			
	8	4					9	

Medium # 457

6						5		
	5	4		3		2		7
			6			3		
		3			2		7	
			5		9			
	7		1			6		
		5			4			
2		6		5		7	8	
		8						2

Medium # 458

					4	3		7
5		3		8			1	
				2				
3		6			8			5
		2				6		
9			7			4		3
				9				
	1			7		9		6
4		5	6					

Medium # 459

				6			1	9
				7			2	
3			2			6		8
6					8		7	
	3						4	
	7		9					6
1		2			6			4
	4			2				
5	8			1				

Medium # 460

		9		4				8
			2	3		6		
		8				3		
	8	3			1	9		7
9		6	7			2	4	
		2				4		
		4		5	3			
6				1		7		

Medium # 461

6						2		4
		9	4					
4	5		3		8		9	
				7	3			6
5			8	1				
	2		5		4		1	7
					7	3		
3		6						8

Medium # 462

	8		7	3				
		3	5		8		7	
9							4	
8	1							
3			4		5			9
							5	1
	3							5
	2		1		3	6		
				5	4		8	

Medium # 463

			8				5	
3		8	4				6	2
					5		3	
	1				9			
		2		4		7		
			1				8	
	6		2					
1	4				8	6		9
	9				6			

Medium # 464

8						2		
	5	3	9				4	
	1	2	3					6
				5			6	
		4				3		
	8			2				
1					8	6	3	
	2				1	7	5	
		5						8

Medium # 465

4				5				8
	5	2						
8		1	3		7			
	4							
1	7	9				6	3	4
							9	
			9		5	4		3
						7	5	
9				8				1

Medium # 466

	8	3						1
		9			6	7		
7			9					
		8			1	9		
	9		6		3		5	
		6	7			2		
					7			5
		2	5			1		
6						3	7	

Medium # 467

	9		4		5			6
	1			9	3		8	2
9				8		5		
		7				1		
		3		5				8
1	4		8	3			2	
7			2		1		5	

Medium # 468

			5	6				
						5		1
				8	1	7	9	3
	6	7	4			3		
		5			9	2	4	
5	7	6	1	9				
1		3						
				2	7			

Medium # 469

								2
				6	3		8	
		9	1					6
	6	5			4			
	8			7			9	
			2			5	7	
3					8	9		
	4		9	1				
8								

Medium # 470

			7					
8				3	4		9	6
6	1	7						
			6	7				4
	9						2	
1				8	9			
						8	4	5
9	6		4	1				3
					3			

Medium # 471

4			5					8
	9		8			2		
2				6				
	7		9				8	
8	4						3	7
	5				4		9	
				2				4
		4			9		1	
1					3			9

Medium # 472

								4
					4	1	9	
			5		6			2
	1			3		7		6
	4	3				8	1	
9		6		8			4	
5			6		7			
	2	9	3					
8								

Medium # 473

			3		9			
	4	3				2		
8				7				6
		5	8			1	7	
6								4
	8	7			5	6		
3				5				8
		2				7	4	
			7		6			

Medium # 474

	2			4				1
			2	8			3	
6		3				5		
		4			7			
	9	1				3	5	
			1			4		
		5				6		3
	6			9	5			
8				6			7	

Medium # 475

								3
2		5	6	4				
			7		8	5	9	
6	3		1					
		8				6		
					2		3	5
	7	1	2		5			
				7	4	2		8
8								

Medium # 476

4			5		6			
		7				4		1
		9		7		3		
7				5	9			
	3						2	
			1	6				3
		2		8		6		
9		1				8		
			2		7			5

Medium # 477

3		2			1			
9				8			6	
					3		7	1
	7	5						
			6		4			
						1	2	
2	3		5					
	8			9				3
			1			6		4

Medium # 478

						2		9
2	5			1	9	8		
8							4	
4			7			6		
			6		2			
		6			8			7
	7							6
		4	8	2			3	5
3		1						

Medium # 479

	4				7	3		
		2			3		1	
			1				5	
6			3	8				9
7								5
5				4	6			1
	9				4			
	6		8			2		
		5	6				8	

Medium # 480

	8		4	9			2	
		3	1					
			2			1	3	6
7							8	
		1				9		
	4							3
4	6	8			5			
					2	6		
	5			7	8		4	

Medium # 481

5			4				6	9
	4					1		
				2		7		
		7			1	5		6
			6		3			
4		1	7			3		
		4		1				
		5					3	
3	9				2			7

Medium # 482

	5							3
8		3		5				
			9		7	2		
3			4		6	9		8
1		6	3		9			7
		2	7		3			
				2		1		4
4							3	

Medium # 483

							4	
4			5		3		8	7
		3			8			5
		6		9				
2								1
				4		7		
1			7			2		
5	2		3		1			6
	9							

Medium # 484

			2	9		5		1
	2	7		4			6	
				8	5			
		1				8		
8								5
		4				2		
			7	6				
	6			1		7	4	
9		8		2	4			

Medium # 485

9	6		4					
					8		3	6
			9	7				
8				9		2		
	9	3				6	8	
		4		2				5
				8	1			
1	8		2					
					7		5	8

Medium # 486

		9				7	2	
	8				2			
3			1		6			
	7					9	5	
8			5		3			6
	9	5					4	
			2		9			7
			4				1	
	4	6				8		

Medium # 487

4		8	1		2			
5		1						
3				8			7	
1				6			4	
		7				9		
	4			9				5
	6			7				9
						1		4
			5		8	3		7

Medium # 488

5			6		1			
				7			1	9
	2		5			6		
2						1	9	
8								5
	9	7						8
		6			9		5	
7	1			3				
			2		6			7

Medium # 489

			9		1			6
9			7					
		4	6			1	5	
		9					2	
1	2						6	3
	5					8		
	6	3			9	7		
					6			5
4			1		3			

Medium # 490

8			5	1				
4						5	8	
			9					2
			6				3	8
	1			7			4	
3	6				4			
7					2			
	2	1						3
				5	7			9

Medium # 491

	4	3				7		
	1		3					
7			5	1		2		
6				4		9	1	
	3	5		9				6
		6		3	8			9
					2		5	
		7				1	8	

Medium # 492

			5				7	2
					8			1
		8		4			9	
	4	7	2	3				
3								5
				6	9	3	2	
	5			2		6		
1			3					
6	9				1			

Medium # 493

7		6						2
			4			5		
		2	7	9				4
		7				8		
8	3						6	1
		9				3		
1				8	3	4		
		8			4			
4						1		7

Medium # 494

		3						
				3		1	2	
	7		8	9		5		
					9		4	
6		8	7		2	3		1
	4		5					
		2		5	3		6	
	9	5		6				
						7		

Medium # 495

	8				4			2
5						6		8
		1		2	8			
				9		7	1	
8								6
	1	9		5				
			9	7		8		
1		4						9
9			1				6	

Medium # 496

3			1		7		9	
		1	6				5	
9				5				
4	7							
	3		9		6		1	
							3	4
				1				5
	2				3	8		
	6		2		8			3

Medium # 497

		5	4			7		
	2	4		5				1
8			2		9			
		6	9					
5								2
					1	6		
			1		8			5
6				4		1	9	
		8			3	4		

Medium # 498

3		8					5	2
				4		1		
2					7			3
7			2					
9				6				4
					3			9
6			5					7
		7		9				
8	9					4		5

Medium # 499

		5			6			2
4	2					9		
				8	9	4		
		9	6		5			
1								3
			3		1	6		
		8	1	3				
		1					4	5
2			5			7		

Medium # 500

	5	2			6			7
6			1					4
					3			
2	9					6		
			7	6	8			
		3					8	5
			8					
1					5			2
5			9			7	6	

Medium # 501

5					3		7	
	9			8			2	
	7	8			6			
						2	1	9
	3						8	
1	8	2						
			1			4	3	
	5			2			9	
	1		4					7

Medium # 502

				5				6
			8	4		1		
8		1					3	9
			6			9	2	
	3						4	
	2	4			8			
7	1					6		5
		6		1	7			
4				9				

Medium # 503

9		4						1
	3				7			2
				4			5	6
2			4	8				
				6	5			7
6	1			3				
4			8				7	
5						8		9

Medium # 504

			5			3		2
4			2			6		
	1				8			
	8		1	2				7
	4						5	
7				6	3		8	
			3				7	
		8			4			6
5		7			1			

Medium # 505

	5	1		6	9			
						9	6	
6				3			5	
		4			5	7	8	
	8	5	9			6		
	3			7				1
	1	8						
			6	4		2	3	

Medium # 506

5	6	7						8
		3		5				
		2				9	6	
			8			6		
		9	5		4	7		
		8			3			
	2	6				1		
				9		8		
1						4	9	7

Medium # 507

		5	8				4	
			9					
		3		4	6			
1		6		7	8			9
5								8
7			3	6		1		5
			6	3		9		
					1			
	4				7	3		

Medium # 508

		2	8		9			
				2				3
6					7	1	4	
	9				4			
	5	6				7	3	
			2				6	
	3	4	7					8
9				3				
			4		8	3		

Medium # 509

	6		3	8				
						1		2
			7		9			3
		2			1		4	
7	5						1	8
	3		8			7		
6			5		8			
3		5						
				2	3		5	

Medium # 510

							1	
		8		2	9		6	
	1			5	6			
	4		6	3				2
2								8
7				8	5		9	
			7	9			2	
	9		8	6		1		
	3							

Medium # 511

		9						7
7				3				6
	3	6	1					5
		1	4					2
			9		5			
8					3	5		
9					4	1	2	
6				2				9
5						8		

Medium # 512

		9			3			
	6	4		9				7
5			7					2
	8			6				3
			8		5			
1				4			6	
3					8			9
7				2		5	3	
			1			6		

Medium # 513

	5		1	6		2		
					4	3		
	8			2				4
8	6					1		7
3		5					4	9
2				3			1	
		1	4					
		4		1	5		9	

Medium # 514

3			6	4		2		
		6		8	7		9	
	1							
1					4	8		
7								2
		8	9					7
							7	
	6		5	7		9		
		9		3	8			6

Medium # 515

						4	7	2
		2					3	
			3					5
		9	4	3			1	
		4	5		2	3		
	2			9	7	6		
9					3			
	1					7		
5	6	8						

Medium # 516

7				3				6
	1						7	
	2	6	4					
	4	7			1			
5			9		4			3
			6			7	2	
					8	2	5	
	7						3	
9				2				7

Medium # 517

		3	8					
					5	7		9
2	6	9		1				5
		7	3					
	4						7	
					2	9		
5				2		1	9	4
6		4	5					
					3	8		

Medium # 518

			9			1		4
5			4			7		
	3					5		2
		2	5				1	
				6				
	7				8	4		
4		8					6	
		1			4			5
9		7			5			

Medium # 519

	1					4		8
	7		4				1	
			5		1	3		
				8		7		6
		4				1		
3		1		9				
		7	8		9			
	9				2		3	
1		6					7	

Medium # 520

5					7			4
			5	6				
	2				1	3		
				2	9		5	
9		2				1		6
	8		4	3				
		8	2				9	
				1	5			
7			8					1

Medium # 521

	6					5		7
	7	9	1	2				
		8			3			
	4				1			9
		7				8		
9			4				1	
			3			1		
				5	8	2	3	
5		2					6	

Medium # 522

	6				1		5	
	9		5	6				7
		1		2				3
	5	7				6		
		3				4	1	
6				7		3		
2				8	6		4	
	8		3				2	

Medium # 523

6		9		4				
	3			6			5	
5			8		7	6		
	1				2	9		
		2	3				4	
		6	4		9			7
	2			8			3	
				2		4		5

Medium # 524

			2				6	8
2						1		7
9			8	6		4		
		8						
			4	3	1			
						2		
		3		9	4			1
6		5						3
4	7				5			

Medium # 525

							4	
9			7				6	
1	3			5	9			
				9		5		
	9	5	6		7	2	1	
		8		1				
			5	6			3	7
	2				1			6
	1							

Medium # 526

5	2			4	7			8
		6	1			4		9
		7			6		9	
	4						7	
	8		9			5		
8		4			2	7		
7			5	1			8	4

Medium # 527

		2			3	6	1	
		1		7			9	5
	7							
	4		8		1			
		5				2		
			4		7		5	
							3	
3	9			8		4		
	8	4	6			1		

Medium # 528

				4	7			2
	4		9					6
5							4	
7					9	6		
		9	4		3	7		
		6	5					1
	2							7
6					5		9	
8			1	3				

Medium # 529

		5			1	6		
1			5		9			4
3							8	
	5					8		
			8	5	2			
		6					3	
	8							2
7			6		3			8
		3	4			7		

Medium # 530

	4		5	2		1		7
			8	9				
					6		3	
	2	8						
5		3				7		6
						5	9	
	1		2					
				4	5			
2		9		6	7		5	

Medium # 531

7		1			8	6		
2				4				1
			2	6				
	5			3			6	
	3						8	
	1			5			4	
				1	3			
6				7				8
		3	6			5		7

Medium # 532

6						9		
			7		3			5
8	1		4				7	
		8			7			
	2			3			4	
			9			5		
	5				1		6	7
2			6		4			
		9						1

Medium # 533

							3	1
6	7			9				
1			7		4	5		
	2							7
5			8		9			4
8							1	
		4	2		5			6
				6			7	3
2	6							

Medium # 534

			8			1		
	4	8		6				9
					2	6	8	
	3	5					4	
	1						9	
	6					8	2	
	8	1	2					
3				9		2	5	
		9			5			

Medium # 535

	5					9	1	
		6	2		5		7	
	9			7				5
8					6	2		
		9	8					4
1				3			5	
	2		6		1	3		
	6	3					8	

Medium # 536

	6				5			
	4	1				8		2
				8				
5		4		9			2	
	9	6				5	3	
	8			7		4		6
				4				
2		3				1	6	
			2				8	

Medium # 537

	7					6		
				6			4	1
		1	4		9			
	4	2		7	1			6
7			5	8		2	3	
			2		7	9		
5	3			4				
		4					6	

Medium # 538

1				4	8			6
	8		5				2	
		9	1					
		7		2			8	
		1				3		
	2			8		7		
					5	6		
	7				6		3	
6			7	1				4

Medium # 539

4				7	6			
	2				1	5		
				2			9	1
	1		3				8	
	6						4	
	3				9		2	
3	4			6				
		7	2				5	
			5	4				3

Medium # 540

						6	1	
			3			7		
9				6	7	2		3
	9				2	1		
5								9
		8	4				6	
3		7	6	2				8
		1			3			
	8	9						

Medium # 541

	4			2	6			
	6		9			3		
			5					4
	5				2			8
7		9				6		1
3			4				2	
8					7			
		7			5		8	
			6	1			5	

Medium # 542

		9		8		7	6	1
8	2							
		5	3					
	7	2						
	9		8		6		7	
						4	5	
					1	6		
							1	2
7	1	4		2		9		

Medium # 543

							1	
	2	9		5		3		
3			6				7	
5			7	8				9
		8				1		
1				6	3			7
	9				7			8
		7		3		6	5	
	8							

Medium # 544

	9		8	3		2		
			7					9
	5	2				6		
		7			9			3
2			1			4		
		9				7	6	
8					2			
		5		1	8		2	

Medium # 545

					8	6	3	9
			4		6			2
				1		8		
	5		3					6
		4				9		
7					2		5	
		8		9				
9			8		7			
6	4	5	2					

Medium # 546

	3					4		
					4			7
	8		2	9	1	5		
		8		3				
	9		1		7		8	
				8		1		
		4	5	6	3		9	
2			9					
		9					3	

Medium # 547

			4			8		
6			7	9				3
	9				1			
		4		3				5
	5	9				1	3	
3				8		2		
			2				6	
9				4	3			7
		7			9			

Medium # 548

	8	3	5					
		9		7		3		
5		2	4			9		
					8			1
	5						9	
7			6					
		5			4	7		9
		6		8		5		
					1	8	3	

Medium # 549

			2					8
7					4	3		
2				3				
	8		5			6		
	9	7	8		1	4	5	
		5			9		7	
				8				5
		2	9					4
1					3			

Medium # 550

	8	5		7				
		6		8	1		7	
9					4			
								8
8		2	5		9	1		7
1								
			2					3
	6		7	3		2		
				9		5	6	

Medium # 551

	6	5					8	
					6			4
			8	2				
2				7		6		
7			1		2			9
		9		6				2
				1	7			
1			9					
	4					3	1	

Medium # 552

					6	2	1	
5	6						8	9
3			7					
	3	6		1		9		
		8		5		1	7	
					4			5
6	1						2	8
	5	3	6					

Medium # 553

7			1		5			2
			4					1
						8		
		5	3			9	2	
1			5		4			8
	2	6			9	4		
		4						
6					2			
2			7		1			3

Medium # 554

4						1		
8	9						6	
				6	1	2		
				5	3			7
		5				9		
7			9	4				
		6	4	8				
	7						9	2
		3						6

Medium # 555

	2			1	7		5	
				6			1	
					2	3		6
6					1			
	5	9				4	8	
			9					7
5		2	4					
	3			8				
	4		1	7			3	

Medium # 556

		6	7					
			3	6		1		5
7	4					9		
	9		1					
	7	3				8	6	
					6		5	
		7					9	3
4		9		5	1			
					3	5		

Medium # 557

	8				3	4		6
		1						
	5		9		6			8
		4			7	1		
			8		5			
		8	3			6		
2			6		8		5	
						8		
8		3	4				2	

Medium # 558

			9		4	7	1	
	6				3			
		9		1			5	
	2		8			5		
6								9
		8			7		2	
	1			4		8		
			3				4	
	3	5	6		2			

Medium # 559

2	4		9					
				3				4
	1	5				8		
		2		1			6	5
3								7
8	6			2		3		
		4				6	5	
6				9				
					6		1	3

Medium # 560

					8		6	
6		8		1			3	
	4		3					7
9			1					4
				7				
8					4			5
1					3		9	
	7			8		5		2
	9		5					

Medium # 561

	5					7		
							8	3
1					8		9	2
		9	8					7
	4			2			3	
6					5	4		
7	3		6					9
9	8							
		4					2	

Medium # 562

	9		1		6			
	4	8		5				9
		1						7
		4		2				
5			8		1			6
				9		4		
1						8		
2				3		6	1	
			9		5		2	

Medium # 563

		1				4		
3					8			1
5	6	9			2			
		2		7				4
	9						3	
8				3		5		
			2			7	4	9
9			4					3
		7				1		

Medium # 564

		9		1	4		8	
							1	
5			7	8	3	6		
	2	6	3					
					9	4	7	
		7	5	9	8			2
	1							
	8		1	2		7		

Medium # 565

			8				7	
9							4	
				4	1		3	8
		5	7	2		4		
		9				2		
		2		1	3	8		
3	4		6	9				
	9							1
	5				2			

Medium # 566

		6	3		9			
					7	4	9	
				2			5	
		7		3		6	2	
		1				8		
	9	8		1		5		
	7			6				
	6	4	5					
			1		3	7		

Medium # 567

9					8	7		
			7	3				6
8				2	9			
	4	5						
			8		6			
						3	4	
			1	6				2
1				7	3			
		7	4					9

Medium # 568

					8		6	7
		4		7	9			
					6	2	9	
3	1					5		
5								8
		8					2	1
	8	5	2					
			1	4		7		
1	2		8					

Medium # 569

	5			4				
		9						1
1	2		9				8	
5	4				9			
8			1		7			6
			5				9	8
	6				3		2	5
3						4		
				6			3	

Medium # 570

								7
7				6	4	2	8	
4		6	3					
		4	9			1		
	3						7	
		8			3	5		
					5	3		6
	2	3	6	7				1
5								

Medium # 571

	9				4	5	2	
		1		7				
			8	1				3
				8			5	
8			3		7			9
	6			9				
4				6	2			
				5		9		
	7	2	9				3	

Medium # 572

			5			1	2	
3			7		4			5
9								7
			3				6	
	6		1		2		7	
	1				7			
4								6
1			6		3			8
	5	6			8			

Medium # 573

	4			9				1
						9		
2		6	7					
	2				4	7		
7			9		6			4
		9	1				2	
					5	1		3
		1						
5				3			6	

Medium # 574

4			6	3				
2							5	
	8	6				4		7
		3			6			2
				8				
5			3			6		
8		4				3	2	
	2							9
				5	7			1

Medium # 575

9					8	4		
5	3							
8			7		2	5	6	
	9			1				
			2		4			
				7			2	
	4	9	8		3			2
							5	4
		1	6					7

Medium # 576

		4						
2			7		6		3	
9			3	8				
	2	5						7
3	6						8	9
7						3	6	
				3	5			6
	5		6		1			3
						2		

Medium # 577

6		7	2					8
			4			5		
	2						9	
		5	3		6			
		2		7		6		
			9		1	8		
	8						3	
		3			2			
7					8	1		6

Medium # 578

2				4	7			
1			2		9			
5	9		3					
	6							7
	7	5				4	8	
3							9	
					2		7	1
			6		5			3
			4	8				6

Medium # 579

			8		9		7	
	8	1		2		9		
			1					2
	9		3	1				
	6						5	
				7	5		3	
5					4			
		6		9		5	8	
	2		6		8			

Medium # 580

	2	1		5		4		
7				4				
			6		7		2	
5	3							8
		6				9		
9							1	7
	9		2		6			
				9				5
		3		8		6	9	

Medium # 581

4	5	8						
		6			9			7
		2	5					
			4	9		3		
8	6						9	1
		3		1	8			
					7	1		
3			1			4		
						8	7	9

Medium # 582

		1	9		5	6		
			4			7	3	
				8				2
					7	8		
	4		1		9		5	
		9	8					
6				4				
	7	2			3			
		3	6		1	9		

Medium # 583

			9					8
2		7		6	5			
1						2		
4	1				6	9		3
7		9	3				5	4
		3						9
			2	9		3		6
9					1			

Medium # 584

	8							
2			1		7		3	8
6		4		9				
5				1		2		
		1				7		
		7		8				9
				4		9		7
9	7		8		3			6
							2	

Medium # 585

					3			9
	1			6		8		
			4	7	1	5		
4			3			7		6
9		3			4			1
		2	5	4	7			
		5		8			4	
6			1					

Medium # 586

3			8	1	6			
			4				5	
		9		2				4
			6				3	8
9								1
1	8				7			
8				6		2		
	7				2			
			9	7	8			3

Medium # 587

	9			3				2
				2	9	3	6	5
							7	
	2	4				7		
			7		3			
		3				8	1	
	8							
6	5	7	1	9				
9				6			2	

Medium # 588

				2				1
7	2							
4			8				3	
	7	5		8		9		3
		2				1		
3		8		5		2	4	
	6				4			5
							9	6
5				7				

Medium # 589

		9	8	6		2		
	4		2					
	5			9	4			1
7		2						
	6						3	
						4		2
9			1	8			6	
					3		4	
		7		4	5	8		

Medium # 590

	7					3		
2				8				
4		9	1	5				
8		1	6				7	
7								3
	9				7	4		1
				2	6	1		7
				4				5
		2					9	

Medium # 591

6		7				3		4
2			5			6		
				3			9	
7						2		
3			6		8			5
		1						6
	1			4				
		6			5			8
5		3				7		9

Medium # 592

	4		9			2		
		5	1				6	
						9	3	5
	3		7		5			2
8			4		9		5	
1	9	8						
	2				3	1		
		6			8		7	

Medium # 593

	6			2				
4		5			1			
		3			7		4	2
				7			1	
8			1		3			5
	3			5				
3	4		7			2		
			9			3		4
				8			9	

Medium # 594

4				5			7	
	5		3		9	8		
7			4					
		6			7	2		
	9						5	
		2	1			3		
					3			2
		5	6		8		4	
	2			9				5

Medium # 595

		8	1					
			7		9	5	4	
7						2		9
			5					8
		9		6		3		
5					7			
3		4						7
	1	6	9		2			
					3	1		

Medium # 596

	8				4	7		
		2	1					
	3			5				6
7	5	8						1
		4				6		
2						9	5	3
6				1			9	
					7	3		
		3	4				6	

Medium # 597

						4		
4	9	5		3			8	
					8	2		6
				7	2	1		
5								2
		1	8	5				
2		4	3					
	3			2		5	9	4
		9						

Medium # 598

7					6			3
	9				3			
6		2	9					5
		9	8			5		6
2		6			7	8		
3					5	6		1
			2				4	
1			7					8

Medium # 599

	8		9			7		
5	6				1	8		
		7	3					9
		2					8	
			1		5			
	3					2		
1					9	3		
		6	5				7	4
		3			7		1	

Medium # 600

9			5	4				
	8	5					6	4
				6		5		
1			4			8		
2								9
		3			7			5
		2		7				
3	9					1	7	
				8	6			2

Medium # 601

					5			
		9			3	8	5	
	6	7		9				
	4					7	1	
6			9		7			5
	9	2					8	
				8		3	2	
	7	8	5			9		
			2					

Medium # 602

7	1	3			4			
				2	9			
4		2						
		1		7				2
3	6						5	8
8				6		4		
						5		3
			6	9				
			8			1	6	4

Medium # 603

	3	8	5	6				9
						7		
	6		2					8
	2			1				
	1		6		3		5	
				4			3	
4					6		7	
		1						
6				3	7	8	9	

Medium # 604

	7		1				8	
		8	9		6	1		
				2		7		
				6				8
	6		4		7		9	
2				5				
		3		9				
		5	3		2	4		
	1				4		6	

Medium # 605

1	9				2	3		
3				4				
	2				7			9
	1	6					9	
	7						4	
	4					1	8	
7			3				5	
				2				7
		5	8				2	4

Medium # 606

							4	5
6	3		1					
5		9	2					
2		7	3					
		8				6		
					6	7		3
					5	8		4
					1		9	7
4	2							

Medium # 607

	6	8	7				1	5
			4					7
	5	1	6					
5	3				4			
			1				6	4
					3	2	7	
9					6			
6	4				8	5	9	

Medium # 608

	3		1	5				
4		9				7	5	
		7			9			
3			2	4			6	
	2			9	8			1
			3			1		
	9	2				5		4
				8	4		7	

Medium # 609

	7							1
8				6			9	
5			3		9			
		5			8	7	3	
1								4
	9	8	2			5		
			5		6			3
	6			3				2
2							7	

Medium # 610

	1				4	2		
	3		2				9	
2					3	7		5
				2				
	7	5				9	2	
				7				
4		9	8					1
	8				2		6	
		7	4				3	

Medium # 611

				7				1
			2		1		3	5
4						8	7	
		8	5					3
	6						9	
1					4	6		
	1	3						2
5	7		9		3			
9				1				

Medium # 612

4			6					
6	5			7	2			
				3				9
	9				3	6		4
	8						2	
7		4	2				3	
8				4				
			7	9			5	8
					5			6

Medium # 613

			2					7
	8					6	5	
7	4							8
	3			9		8	7	
			6		3			
	1	2		4			6	
9							3	1
	6	3					4	
5					9			

Medium # 614

	5		2	6				
		9			3		4	7
6	7							
5	2			9				
7								6
				7			3	4
							1	5
9	4		5			3		
				4	2		9	

Medium # 615

	4	3			8			9
			7					
	9	6						5
	1			9				2
		4	2		5	8		
2				4			1	
4						6	8	
					3			
5			4			2	9	

Medium # 616

			9	1				
8		3				9	2	
		7	4				6	
2	6							
		9		3		6		
							8	2
	7				1	8		
	4	6				1		9
				7	9			

Medium # 617

	3			6		8		
			1	9				7
	1					4	6	
9	6				7			
			4		9			
			8				4	3
	9	6					8	
4				5	2			
		5		8			2	

Medium # 618

	8		4					9
	4			1		7		
9			2		3		4	
			3		8		9	
	1		9		7			
	6		7		2			5
		7		8			6	
3					6		2	

Medium # 619

		4	2				3	
5								2
7	8		6					
		6		5				8
		9	8		4	2		
1				7		3		
					9		2	3
6								5
	1				8	6		

Medium # 620

			4	2				6
			8			9		1
	2				7			3
					5	3	9	
	8						7	
	5	3	7					
4			9				1	
1		7			4			
8				6	3			

Medium # 621

		7			1			
8					9		6	
	6	1					4	
5				8		7		
2			9		7			6
		8		5				1
	4					3	1	
	2		3					9
			6			2		

Medium # 622

	7			6		5		
	6		4	7		8		
			5			6		
2		8		5		4		
		1		2		7		3
		6			7			
		7		1	3		4	
		2		4			1	

Medium # 623

					5		8	
			2	3			4	
		9	4			7		
		2	9					5
		7	8		6	9		
4					7	3		
		4			2	1		
	7			1	9			
	6		5					

Medium # 624

		8		1				
2			4					
6		5			9			2
	3			4				
9	2		8		1		5	6
				5			2	
8			7			4		3
					4			5
				8		6		

Medium # 625

	2		3	8				4
							3	
3				2				5
8			4			6		
		9				1		
		7			6			9
4				3				1
	8							
1				4	2		5	

Medium # 626

								1
1		5		9		6		
		4	8		5			
2	7				8		3	
			7		2			
	3		5				4	7
			6		1	5		
		3		8		7		2
6								

Medium # 627

			2			9	6	
9	2					5		
	6						3	2
6			4					
		4				2		
					5			8
8	4						9	
		7					1	3
	5	3			6			

Medium # 628

1		3				9		
				4			7	5
		2						6
			6		5		2	
	1						8	
	4		2		8			
9						8		
6	2			3				
		5				4		2

Medium # 629

			3		1			9
				7	2			3
4		7						1
	4		2			1		
			5		9			
		6			8		4	
7						5		2
6			7	5				
1			9		4			

Medium # 630

5	6		1		3			8
							6	
	4		5	2			1	
			2					
		5	3		1	7		
					8			
	9			5	2		4	
	1							
8			9		4		2	3

Medium # 631

				9				1
		7	4		1		6	
		1				4		7
7			3					6
	2						8	
6					8			3
5		4				8		
	7		9		2	5		
1				7				

Medium # 632

1		4	3	5				
		8	9			6		
9	3						1	
					5	8		
		6				3		
		9	7					
	4						3	8
		5			7	2		
				9	1	4		5

Medium # 633

	2		9		8			
				5			7	
	8	7				3		
				6	5	7	2	
6								3
	3	2	1	8				
		1				6	5	
	4			2				
			5		9		4	

Medium # 634

3		1					6	
	6			7			4	
8			6					2
		6	1	4				
	9						7	
				2	3	1		
4					9			5
	7			1			8	
	5					6		1

Medium # 635

	7	4	9			5		
							2	
		9		8			1	
7		5		1				
9			8		3			6
				7		4		3
	5			3		7		
	8							
		7			4	2	8	

Medium # 636

		4			1		3	
		5		4			9	
6				5	2	4		
		9			3			
1								3
			5			6		
		7	6	1				4
	9			8		7		
	6		2			8		

Medium # 637

	3				5		6	
1		9						
4	8				2			9
9		5		1				
			3		7			
				6		8		7
5			6				7	8
						1		6
	4		8				3	

Medium # 638

				4			5	
4		2		7				
	7		9	3		2		
9							8	
7		8				1		4
	5							9
		7		8	2		3	
				1		7		6
	1			9				

Medium # 639

9							7	
			6	2				
					3	6	9	5
		8		6		5		
	1			5			6	
		2		3		7		
1	4	9	3					
				1	8			
	5							6

Medium # 640

					7	4		
				8	4		7	6
						8	9	5
				1		9	3	
	7						6	
	1	3		2				
7	6	5						
4	2		9	3				
		9	2					

Medium # 641

			7			9		3
	1		4		2			
	7						2	1
			8			5		9
2								7
9		1			3			
6	8						1	
			1		6		8	
1		3			4			

Medium # 642

4	9	2				7	5	
			4	5				8
8	5		6			1		
		4				6		
		9			2		3	5
2				6	1			
	4	3				8	9	6

Medium # 643

	6							
			4	2				
						1	3	9
7		8	3	9		6		
9								1
		3		1	7	8		4
6	2	9						
				8	3			
							5	

Medium # 644

	5		1					8
8	7							
	1			2				7
		7	6			3		
	4			3			7	
		6			4	5		
9				8			3	
							2	1
5					6		9	

Medium # 645

		2			8			6
1		5						9
	4			7			3	
		8	9					
7				3				5
					1	7		
	3			1			9	
4						8		1
6			4			3		

Medium # 646

5				1	3			
	2			5		6		
9		8			2			
							7	
	9	2	8		7	5	3	
	5							
			3			2		7
		6	2				4	
			7	6				3

Medium # 647

			9				2	
					3		7	6
		6			1			9
						7	8	
6		3				2		1
	7	5						
7			4			8		
2	9		5					
	3				9			

Medium # 648

			2				8	
9		1	6	7				
		4		8				
3		9						
	7	2		3		4	9	
						8		5
				1		7		
				4	7	5		1
	4				3			

Medium # 649

			9			5		6
8					2	3		
2	1						7	
				4	7	6		
	4						1	
		8	1	3				
	5						8	2
		2	8					3
9		7			4			

Medium # 650

4	7							9
	3					8	7	
6				8		1		3
			1		9	4		
		3	4		2			
3		5		6				1
	8	7					5	
9							3	7

Medium # 651

	3							
4	1		5					
5		9	2	6				
				4		6	7	9
3								2
2	4	6		9				
				7	6	3		8
					1		2	5
							9	

Medium # 652

							1	9
	9	8	6					
6			5			4		
			4	6	1			2
1								3
2			3	8	5			
		2			4			1
					7	5	9	
7	3							

Medium # 653

7		9				2	3	
				9	4		8	
			6					
		2	7				4	5
				5				
3	9				6	7		
					7			
	2		9	1				
	6	7				9		1

Medium # 654

		3	8	9			2	
1					7			
	8			2			9	
6			1	7		5		
		8		6	5			7
	9			4			5	
			5					6
	4			8	6	9		

Medium # 655

8			7			4		2
		7	9	8				6
4								
		1		3			8	
9								1
	3			5		6		
								4
5				9	8	2		
2		8			6			5

Medium # 656

		6	7			4		
	1		8	5				3
			9				7	2
			3			5		
3								9
		4			1			
4	3				5			
8				9	2		3	
		2			7	9		

Medium # 657

						7		1
1					8	3		
			6			2		
7		2	9	3			8	
	3						7	
	6			4	7	9		5
		5			2			
		8	3					9
2		9						

Medium # 658

		1		4			9	
		4				2		
6			8	1				
		9	3	5				7
2								3
3				6	7	5		
				7	4			5
		5				6		
	4			8		9		

Medium # 659

2	6		7			8		
	9			8				4
		3			4			
		9		7			4	6
1	4			2		3		
			8			1		
4				9			6	
		1			6		2	9

Medium # 660

		2		3				8
4				2			6	9
		1	5		9			
1			3					
		6				2		
					7			4
			2		3	7		
6	3			9				2
2				5		3		

Medium # 661

		6	8			3	5	
4				3			7	
		5		9				4
	4		9				1	
	2				5		8	
6				8		1		
	9			6				5
	3	7			2	8		

Medium # 662

	2		6					
			1					
5		1		3	8		9	
8	7			9		1		3
4		9		7			8	5
	8		2	1		4		7
					6			
					5		3	

Medium # 663

5		2			9		6	
1				8				
3		4						1
	1	5						
6			3		5			7
						5	4	
8						9		4
				9				3
	3		7			6		5

Medium # 664

		3		4		2		1
			5	7				8
	1							
3	5						4	
			1		9			
	7						5	6
							9	
2				9	3			
6		5		2		7		

Medium # 665

		5					1	
		8	6	4				
6					3	2	5	
				6	8		2	
		2				8		
	8		7	3				
	2	4	8					3
				5	4	6		
	7					1		

Medium # 666

5			2					1
	7					8	9	
	8			6		5		
					2	3		
	3		1		5		6	
		8	7					
		5		9			2	
	2	4					3	
9					1			8

Medium # 667

		8	3		9		2	
1							7	
	6		7					3
						2	5	4
		9				7		
4	5	2						
6					7		4	
	2							6
	8		9		6	3		

Medium # 668

1			6					
2				9		4	8	
	5					9		1
		8	1	3	2			
			4	7	8	1		
5		4					7	
	7	3		5				4
					6			2

Medium # 669

	8			3				7
				8		9	5	
3			6					
	7	5					6	2
2								5
4	1					3	9	
					5			9
	3	9		4				
7				2			1	

Medium # 670

						3	7	
	8		6	7				
7		2		3				
		4		1	7		5	8
9	2		3	6		4		
				4		6		5
				8	6		2	
	7	9						

Medium # 671

5								
		3	4		1		9	
9				7	3		1	
		4		1				
7			6		9			8
				5		4		
	2		7	3				5
	9		2		4	6		
								4

Medium # 672

		3			1	2		
		8			3		7	
			2				9	
	6		5			1		
9		7				5		6
		2			6		8	
	1				2			
	2		8			4		
		5	3			9		

HARD

Puzzles

Hard # 673

	8			9				
5		9	1					8
		4	7					2
			5				2	1
	7						6	
3	5				9			
2					7	9		
9					1	2		5
				3			1	

Hard # 674

				7		3		8
	3				2		9	7
		8						
7				1		9	2	
6								4
	1	5		9				6
						7		
9	5		3				8	
4		3		5				

Hard # 675

1	7						4	
					5			9
8			9	1				
	4	6	5					
		8		7		4		
					1	3	7	
				4	9			1
2			3					
	3						5	7

Hard # 676

4			5	1			7	
							9	4
	2				4			
3	6			2	7			
		8				3		
			6	3			8	5
			9				6	
2	8							
	7			8	2			1

Hard # 677

	8		5	7				3
7								
	6	1		3			8	
		4						
3		6		4		1		7
						4		
	4			2		9	1	
								2
1				8	3		4	

Hard # 678

7	1		6					
			3					5
						4	9	
				2		8	7	
8		7		4		3		6
	6	9		8				
	7	8						
5					1			
					2		5	4

Hard # 679

6	7			2				9
2						8		
		8	3		9			
	2			1			5	
			8		2			
	8			7			1	
			4		5	7		
		9						5
1				8			3	4

Hard # 680

7				5				
4		1				8		
			9	8	1			7
				1	9			6
	2						5	
8			2	3				
6			8	2	3			
		2				1		8
				7				3

Hard # 681

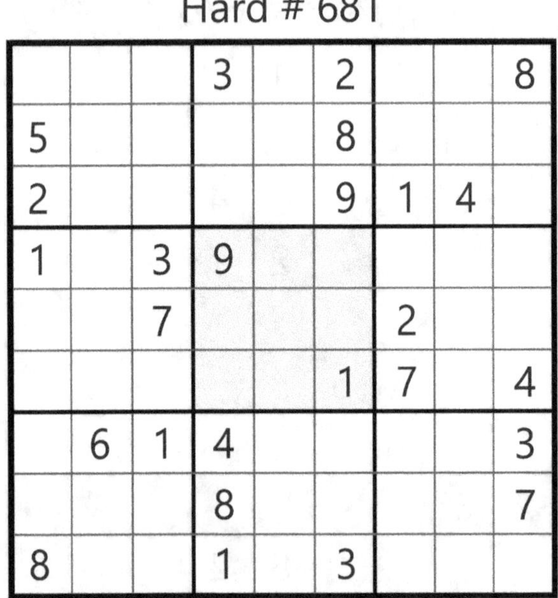

			3		2			8
5					8			
2					9	1	4	
1		3	9					
		7				2		
					1	7		4
	6	1	4					3
			8					7
8			1		3			

Hard # 682

		8		2	9			
2								
	7	5	6					
				6		7		1
1			2		3			4
3		6		9				
					8	4	1	
								5
			5	7		6		

Hard # 683

8			6	4				2
	2					4		
					2		1	3
3								1
		6		7		8		
1								6
6	5		2					
		9					5	
7				8	5			9

Hard # 684

				8	9	2		
				2		3		
2				5		1		4
5			4					
	4	9				5	8	
					5			7
7		8		3				1
		5		4				
		6	8	9				

Hard # 685

	8						1	7
						6		
			4		6	9		
	9		1	2				3
2								8
3				4	5		9	
		4	6		9			
		8						
7	2						5	

Hard # 686

			9					
2					5	4		9
5		7		3				
		2						4
	8	3		6		7	5	
7						8		
				7		9		6
6		8	5					1
					6			

Hard # 687

		6	1	2	8			
							8	
	4	8	7					
1		7		4			9	
	9						7	
	2			6		5		1
					4	9	2	
	5							
			3	8	5	7		

Hard # 688

	5	4		3				
			6	9				
6	7						8	
		5			3		2	
	9			8			4	
	6		2			1		
	3						5	8
				2	7			
				1		2	9	

Hard # 689

		8			6			2
1				8				
		3			1		8	
	5				7	9		
6			1		9			8
		9	3				4	
	1		9			5		
				1				4
2			6			7		

Hard # 690

9			1					
			8		5			6
	7		2			4	5	
7	4					3		
		2				9		
		1					4	5
	8	4			2		1	
5			9		8			
					1			4

Hard # 691

			7					9
		9						2
6	3			9	8		7	
				2			8	
		5	6		7	4		
	1			4				
	2		5	8			3	6
5						1		
7					2			

Hard # 692

	3	6	5		7			
4		7		2				
1					6			
		2		8				
	8		6		2		5	
				5		9		
			2					1
				1		5		9
			4		8	7	3	

Hard # 693

2						9	6	
			8			7		
		1						4
3		4			6			5
	2			9			4	
6			5			2		7
4						5		
		6			8			
	7	3						1

Hard # 694

	7	1				6		
	3		9	7				
		2	4					
3					1	9		6
	9						3	
6		8	3					2
					7	2		
				2	3		4	
		9				8	1	

Hard # 695

8								
				2	5		4	
	1	3						
	5	4		8		3	1	
	7			4			6	
	3	6		7		9	8	
						4	9	
	8		6	5				
								2

Hard # 696

					4		1	
6	9			1				3
	3		9				8	
		9	3					7
			7		6			
4					5	3		
	8				2		6	
2				7			4	9
	4		1					

Hard # 697

				4			9	
	9					5		
		5	1		9		8	
8		1	3		2			
7								3
			4		6	7		1
	5		6		3	1		
		2					5	
	7			5				

Hard # 698

		5			7		4	
			4	6		7		
6						3		9
			1	2				8
9								2
1				3	6			
7		3						1
		9		4	2			
	6		8			9		

Hard # 699

4	3			2				
		8	9				7	
					7	2	6	
9				6	3	5		
		1	2	8				9
	7	2	3					
	4				8	7		
				1			5	6

Hard # 700

2	8		3				5	
		1		5				
		4		8	7			
1		9			3			
4								6
			2			3		1
			7	3		6		
				2		8		
	5				8		2	4

Hard # 701

		9	8	7			4	
				2		6		
5	6		9					
		3						
7	4	8				5	1	6
						2		
					8		6	4
		7		6				
	8			4	1	7		

Hard # 702

					6			4
3			7				9	
	9	6					3	1
			2					
5	4						1	9
					8			
9	5					3	8	
	7				3			2
1			6					

Hard # 703

9			5				4	
		2		3			1	
						6		
			9				7	8
8		7				4		2
2	9				3			
		6						
	7			9		2		
	1				4			6

Hard # 704

			6	7				
			1			9		6
	2						4	
4	9				8			2
		5				7		
8			5				6	9
	5						9	
7		3			5			
				8	4			

Hard # 705

		2	1					8
	6	9	4					
5	1					4		
			9	3				7
9								1
4				6	1			
		7					3	4
					5	7	6	
2					8	5		

Hard # 706

				1			2	6
7						3		4
		6			4	5		
				5	8		9	7
9	4		3	7				
		9	4			7		
5		7						8
6	3			2				

Hard # 707

5				7			8	
		7		4	1			
6			8					7
			4		8	5		
		1				2		
		6	5		9			
3					4			5
			3	9		7		
	2			6				4

Hard # 708

					7			
9			6	4			2	3
	6	3						
	5		2				1	
		7	4		8	9		
	9				3		6	
						1	5	
6	3			5	2			9
			9					

Hard # 709

		3		2			8	9
				8		1		
2			7		4			
	2			9		4		5
6		4		3			1	
			6		3			1
		1		7				
9	3			4		8		

Hard # 710

		7	3					
					9		7	3
3	8			6				
				3	5		2	
		5	1		2	3		
	7		9	8				
				2			1	7
5	9		4					
					3	6		

Hard # 711

				3				6
3	8		1			7		
		6			7	8		
				4				5
	3						6	
2				1				
		1	4			2		
		2			6		9	4
7				2				

Hard # 712

	4			1				2
	3	8			5			
			2		8			
1						5	7	
		7		3		9		
	5	6						4
			7		9			
			8			7	1	
3				2			8	

Hard # 713

6		7	1					8
					2		9	
		2		7			1	
			2	8				6
	3						7	
1				3	7			
	1			4		9		
	5		6					
9					3	7		4

Hard # 714

					9		6	
	6	9					5	
2		7				9		
	9			8	7			
8			6		3			4
			4	1			3	
		8				7		3
	1					5	9	
	2		9					

Hard # 715

3								5
		7	5				9	3
	9	6				1		
				3		2		
		9	1		8	6		
		2		6				
		3				8	2	
2	1				4	3		
7								9

Hard # 716

	5		3	1		8		
9				4	8	6	7	
5			7				3	
		3				2		
	8				1			5
	7	5	1	8				9
		4		2	6		8	

Hard # 717

		6		1				5
					7			2
		5			2	9		8
		8				2	5	
				5				
	4	9				8		
7		3	9			4		
8			3					
9				7		6		

Hard # 718

	8		6	9	3			
2			4					
		3					1	
8		5			4			6
	4						3	
9			7			1		5
	5					2		
					9			8
			3	2	6		9	

Hard # 719

			1	9	5			
9						6	4	
		1						
	6	5			2			
		9		1		4		
			6			5	7	
						3		
	7	4						5
			7	3	9			

Hard # 720

9				4				
			1	2		9		
7	3				9	6		
	5				1			6
		3				4		
1			8				2	
		2	9				8	4
		5		6	8			
				5				9

Hard # 721

7					6	4		
							9	
9			1		2	8		
3	5							4
		1		5		6		
6							3	1
		3	4		5			7
	2							
		4	6					2

Hard # 722

			8			6		
	9		4		5			7
		6		9			5	
	2	4			3			
	5						1	
			5			4	3	
	3			8		1		
5			9		2		7	
		7			4			

Hard # 723

	5				2	7		
	6			4			8	
		4			8		1	
			4				5	3
		7				1		
4	2				3			
	1		5			4		
	8			9			3	
		9	2				7	

Hard # 724

	7	4			1			
3		1					6	
8	9		5					
		9		2	6	3		
		2	7	4		5		
					7		3	1
	4					7		9
			1			4	5	

Hard # 725

4		5			7			
				3			7	1
9					6	2		
	5						1	9
6								4
1	9						8	
		9	4					8
8	2			5				
			1			9		3

Hard # 726

4			3					
		5			2		3	
3	8							6
6			7			2		
		1		4		6		
		2			9			5
1							9	4
	6		9			5		
					1			8

Hard # 727

			8		6			3
	9							
1	7				2			9
	4					6		
9			7	8	4			1
		7					9	
7			1				5	8
							3	
2			6		5			

Hard # 728

	3				1			5
5			4			9		
		1		2			4	
3		8	9		7			
			8		5	1		6
	5			9		2		
		3			4			9
6			3				7	

Hard # 729

2			3			1		
		8	1		2			7
				5			9	
6		4						8
	7						4	
5						3		2
	3			2				
1			4		7	8		
		7			8			5

Hard # 730

	3			2		9		
	1	9						2
			7		3			
		4		5	1			
		1		3		5		
			6	4		8		
			8		9			
6						3	4	
		2		7			5	

Hard # 731

		9		1		6		
1				4			3	5
					7	4		
9	7				8			
		4				5		
			2				9	8
		3	7					
8	9			3				6
		5		2		3		

Hard # 732

	9				5		8	
5					7	2		
		1					3	
				7	3			1
		2				4		
7			6	9				
	3					6		
		5	9					4
	2		7				1	

Hard # 733

2	9				7	4		
	5							3
					9		1	
					1	7		
	8	9	7		6	5	2	
		6	4					
	1		2					
9							4	
		3	5				8	1

Hard # 734

			2	7				
		3		6		2	4	
	6		4			7		1
					4		8	
1								2
	4		8					
5		7			2		9	
	9	6		5		3		
				3	9			

Hard # 735

		2		3		8		
	3					6		9
			2	6	4			
	5	4						7
			7		6			
9						5	1	
			6	9	2			
3		5					8	
		9		8		2		

Hard # 736

			6	5			2	
		9					8	
			7		2	1		5
	7					8		
		8	5		1	6		
		3					1	
3		2	9		7			
	1					9		
	4			3	6			

Hard # 737

	1	2						7
			2				9	
	8				3			
	7	4			5			1
			9		4			
8			1			7	2	
			6				4	
	3				2			
4						8	1	

Hard # 738

8						3		
		7	2				5	
				3	4		7	
		4		1				2
	7	9				6	4	
3				5		7		
	2		9	7				
	4				5	8		
		3						7

Hard # 739

					6			
				8	3	9	6	
	2		7			5		
1		7	8					
4	9						3	8
					2	4		7
		3			4		7	
	5	4	9	3				
			6					

Hard # 740

	6				9	8		
		2	6			4		9
				5				
1	7							4
6				8				5
8							1	2
				2				
3		8			1	9		
		7	9				6	

Hard # 741

	7			9		5	6	
	1		5					2
				3	8			
		4		1				
2	3						1	9
				5		6		
			9	7				
4					1		8	
	6	7		8			2	

Hard # 742

	7	3	5				8	
			1			4		7
						6		
9	2				1			
	6	8				2	1	
			6				5	9
		2						
1		5			3			
	8				5	9	6	

Hard # 743

5			3					6
1		8	9					3
	4							
		3	7			1	8	5
7	5	2			8	6		
							4	
6					1	2		8
8					2			9

Hard # 744

4	8	1			9			
					3			
9			8		2	4		
		2					8	4
	5						1	
1	6					5		
		5	2		4			1
			9					
			3			7	2	5

Hard # 745

		6					2	
	8				7			6
1			5	8				
			6		1	9	5	
4								7
	9	1	7		3			
				7	5			8
6			9				3	
	7					1		

Hard # 746

1			9					2
		6			2		3	
9				5			6	
2	8		6	1				
				3	7		8	5
	1			8				4
	4		1			2		
6					5			3

Hard # 747

					3			
		9	1	4				
	2		6			8		3
4		1		7				2
	6						7	
9				5		3		6
6		4			2		3	
				6	5	1		
			8					

Hard # 748

			8					1
		3						7
7	5		2	3				
	7	4						6
		1		2		9		
3						5	7	
				7	4		9	5
8						6		
4					9			

Hard # 749

3	7			8		9	1	
	8	6						
			4		7	6		
					8			
	6	7				3	4	
			9					
		2	8		1			
						7	2	
	1	4		6			5	3

Hard # 750

				6	3			
7		9						
1			9		2			4
5	2					8		
	9						2	
		7					1	3
4			2		5			1
						6		7
			1	8				

Hard # 751

4	9				1		3	
		6	3					
	2		4			9		5
9				3			1	
	3			8				2
3		7			4		2	
					5	7		
	1		6				5	8

Hard # 752

						6	3	
	6			9			7	
9	8		4	7				
2							6	
8				2				5
	4							3
				5	3		9	6
	7			8			2	
	9	8						

Hard # 753

		6		7			2	1
			1	4			7	
	7							
	3	7	2	6		5		
		4		8	5	6	1	
							4	
	6			3	9			
9	4			1		8		

Hard # 754

5	1						4	
			5	2				9
			3		9			8
		8		5			9	
			8		6			
	6			4		7		
1			4		5			
6				9	8			
	2						5	7

Hard # 755

			8		5	2		
3	9							
		5					4	
				6	2	8		
9	6			7			3	1
		7	4	3				
	5					6		
							2	4
		3	9		4			

Hard # 756

	2	8				3		5
						7		
	1			8	2			
	3		1	6				
		9	2		5	4		
				3	9		6	
			7	5			3	
		4						
6		1				2	5	

Hard # 757

7				2				
			6				8	1
			3	8		6		
6	5		9					
	9	3				4	2	
					3		9	6
		5		3	8			
2	8				7			
				6				4

Hard # 758

	9			4	2			1
7	1				8			
				9		3		
3	4					2		
				7				
		9					1	4
		4		3				
			1				3	6
8			5	2			9	

Hard # 759

1			4					7
	9					5		3
	8		5					
		8			3	9	6	
	6	2	1			4		
					2		3	
8		7					5	
5					4			9

Hard # 760

	3				2			6
			4					7
		8	7	6				3
	2	9		5				
		6				1		
				2		5	3	
1				7	4	8		
4					6			
6			1				9	

Hard # 761

				2	5			3
9			7	6				
						5	9	6
				8				4
		7	6		1	8		
5				7				
6	4	9						
				5	6			7
7			2	1				

Hard # 762

					6	8		
					4		9	
3		1					7	
4			2			6		
1		9				7		8
		8			7			5
	7					1		3
	4		9					
		3	4					

Hard # 763

	1	7	2					
4			3				5	
5			4					
7	2					8		3
		5				2		
9		8					4	7
					6			2
	7				1			6
					3	9	8	

Hard # 764

		9			2		6	
		4	6	8				
		3	1					2
2		5		9			7	
	8			2		9		4
3					5	8		
				7	3	4		
	5		2			1		

Hard # 765

								2
7					3	8		
		6	5	1				
	9			7		5	8	
		8		4		6		
	5	1		2			7	
				3	1	4		
		2	7					6
4								

Hard # 766

8		5	4					6
					7		1	
			6	5		2		
7				3				
4	9						6	3
				8				9
		1		6	2			
	4		8					
6					3	8		1

Hard # 767

					2			6
4		1	5			3		
		3		9				
	4				5			9
2			3		6			1
5			7				8	
				5		2		
		5			3	7		4
6			8					

Hard # 768

7			4	2	1		9	
		5			3			
	6			5				
		6				8	2	5
8	3	9				6		
				4			5	
			2			9		
	9		1	8	6			7

Hard # 769

3		6	1				9	
		5			8			
	9		7	4				
				5	1	2	8	
	6	2	3	8				
				7	4		6	
			5			9		
	5				3	7		4

Hard # 770

	1							
3			4	1				
6		8		7			5	
		5					7	4
			6		8			
4	9					1		
	8			2		3		7
				3	1			2
							8	

Hard # 771

		5			6			7
	2	9		8				
							1	2
4					5		6	
		6	3		2	8		
	3		8					5
6	5							
				1		7	2	
9			6			5		

Hard # 772

	3		2	6	5			
		2		1				
1	8					5		
8	1						4	
				3				
	6						5	7
		9					8	4
				7		2		
			1	9	2		3	

Hard # 773

		6		9				
		1	4	7	8			
5					2	1		
							7	5
8	3						1	2
9	1							
		3	5					6
			2	1	9	8		
				6		4		

Hard # 774

9			3		7		5	
	7				8			1
	1						2	
7				5				
		6				4		
				2				3
	4						6	
2			1				9	
	8		5		3			7

Hard # 775

8		6		4		7		
				6	1			
	9					2		
			6			1	3	8
	5						2	
3	2	8			7			
		7					9	
			5	8				
		5		9		8		3

Hard # 776

9		5				4		1
	3			1	2		8	
			3			9		7
		9		8		2		
7		1			9			
	5		6	9			4	
4		2				5		8

Hard # 777

				3			8	
					8	4		3
	3			9		7	1	
					3			2
6	5						3	8
9			6					
	1	4		5			2	
8		5	1					
	9			2				

Hard # 778

		9		7			6	
	7				3			9
	8	6				4		
			3				1	
			4		6			
	5				2			
		7				8	9	
5			1				7	
	2			9		1		

Hard # 779

1			2					8
					3	6		
9				5		7		
	6				1	8	9	5
5	7	9	8				1	
		7		3				2
		1	5					
8					9			7

Hard # 780

			7		6	1	2	
				2				
	3		5					9
	5						7	
6		7		8		3		4
	2						5	
2					8		1	
				1				
	9	5	6		4			

Hard # 781

		1				9		4
		9		5	8	3		
	8						1	
			3		6		2	
1								6
	6		4		1			
	3						9	
		8	2	7		1		
2		5				7		

Hard # 782

7		3						8
		6	4					
4					9	2		1
		4		5			2	
			9		3			
	3			2		9		
1		8	3					4
					7	8		
5						1		9

Hard # 783

8	1				7		6	4
6							3	
							2	1
				6	4			
	6		3		2		7	
			5	9				
1	8							
	3							9
9	7		2				1	3

Hard # 784

		3	5			2		
		8						
			8	9	3		7	
8			3				6	
3	6						9	2
	9				6			1
	5		9	2	7			
						1		
		2			1	5		

Hard # 785

			8					
	1			5		8	2	3
					3	1		
9		7						6
			3		7			
2						9		5
		3	6					
1	4	8		2			3	
					1			

Hard # 786

	3	8		7				
	9		1			7		3
		6			4			
		2		4			9	
	5						1	
	6			3		4		
			8			6		
1		3			5		4	
				9		8	5	

Hard # 787

		2		1				8
7	3					5	6	
		9	5					
	7				8	1		4
1		8	2				7	
					4	3		
	6	7					2	9
3				9		8		

Hard # 788

	5				9			
			3	2		7		
			1	8		6	3	
	7			3		8		2
8		5		9			1	
	2	3		6	4			
		9		5	3			
			8				4	

Hard # 789

8		4		5		3		
				3	9			
			7				2	
3	5			7				
		1	2		3	8		
				9			7	3
	6				7			
			9	8				
		2		6		4		1

Hard # 790

	1	9		3				
			7		2		4	
	6	7						
1			2			9		
		4				2		
		8			6			1
						5	7	
	4		8		3			
				6		3	9	

Hard # 791

		1	8					3
9	4							
		8			6	5		2
7					8		9	
			9		7			
	6		2					5
5		3	7			1		
							3	7
1					2	6		

Hard # 792

		7	9				1	
	6			7	5	4		
3							7	
					6	1	9	2
2	3	4	8					
	9							7
		3	6	9			5	
	5				8	3		

Hard # 793

			3		4	5		
1	5							
		7		1				3
3	4		8		5			
			4		7		3	6
2				6		4		
							7	1
		8	7		9			

Hard # 794

				1		6		5
	1	2						8
			7		6	4		
	7		9					
		5				1		
					5		2	
		6	4		3			
2						8	6	
3		8		6				

Hard # 795

	5					6		
				4				3
		3			1			5
		7	8	2			4	
		8	5		6	9		
	6			1	4	5		
8			1			2		
2				3				
		4					9	

Hard # 796

5		3		8	9		1	
	9	6						
	2					5		4
			5	7				
3								1
				4	6			
9		2					6	
						4	7	
	4		6	5		9		2

Hard # 797

	9			2		7	4	
		3		1				
1	8				4	6		
					1			4
				3				
5			8					
		1	7				3	2
				5		8		
	7	4		8			9	

Hard # 798

	4		8					
	3	1	7				2	
					6	5		
7					3		8	4
4	6		9					2
		2	6					
	8				2	9	3	
					7		5	

Hard # 799

		4				1		7
2			1			8		4
						2	3	
	9		4		3			
	5						7	
			6		1		2	
	7	5						
3		6			8			9
1		8				3		

Hard # 800

8	1			7				
2	7			4	6			
		9					7	
			6			1		9
1								3
3		8			5			
	2					3		
			5	1			9	7
				3			5	8

Hard # 801

	4	1	7					
		3			8			2
					5		4	
7				4			6	3
	3						9	
2	9			5				1
	7		6					
3			8			6		
					4	1	7	

Hard # 802

				7	6			9
6		9			5		8	
	4			1				
		5				9	4	
3								8
	2	7				6		
				9			2	
	8		6			7		5
2			4	8				

Hard # 803

	3		6	4	9	5		
								3
8		4				7		
7			9					
6			5		3			2
					1			6
		1				9		7
3								
		6	7	3	2		1	

Hard # 804

1		3	6					
	7			2				4
			5			8		
		1	9				4	
	4	6				7	3	
	2				5	9		
		2			8			
7				1			5	
					4	6		7

Hard # 805

			3			5		
2		1					6	
				2	4		1	3
7				3			2	
			9		5			
	6			7				5
9	5		8	4				
	7					2		6
		8			7			

Hard # 806

5		8	6	2				
	2						7	
4				8	9	2		
	4	2						
1								3
						4	9	
		6	1	4				7
	1						5	
				9	3	1		2

Hard # 807

	9		8		1		6	
2	1	6		3				
						2		
		5		4			9	
7								3
	3			9		6		
		1						
				8		5	1	2
	4		5		7		3	

Hard # 808

	2				6			
3						2	1	
		8	4	7				
	9			2	3	8		
			7		8			
		2	9	5			6	
				6	4	7		
	7	1						5
			3				4	

Hard # 809

		3			4	9		
			7			1		2
8		7		6				
	3			5		4		
			8		1			
		6		4			8	
				9		7		5
9		4			2			
		1	6			2		

Hard # 810

7	5			3			8	4
				5	9			
		9						
4		5			8			1
	2						5	
1			3			8		2
						1		
			8	7				
2	1			4			9	8

Hard # 811

2					5	7		
			4		7			
	7	5			6			3
5							1	
	9	8				2	3	
	1							7
7			6			1	4	
			3		9			
		9	8					6

Hard # 812

	9	4	3				6	
			5					9
		1						2
			8			3		5
			6		4			
7		5			3			
1						7		
6					8			
	2				1	9	8	

Hard # 813

		9			2	3		
2				4				5
	3				9		4	
5	9				8		6	
	4		7				2	3
	7		6				1	
1				5				6
		8	3			4		

Hard # 814

	1			2		8		
				9				5
	6					4		1
		3			8	1		2
			7		9			
5		8	2			3		
3		7					1	
1				8				
		6		5			2	

Hard # 815

		7			5	6		
5		4						
	9				8	1		5
	8	2			7			
			9		2			
			6			5	8	
3		9	2				5	
						2		3
		8	1			9		

Hard # 816

				4	2		6	
4		3	7	1				
						9		
1			6					3
		8				7		
7					4			2
		2						
				7	6	5		8
	1		8	5				

Hard # 817

	9		1					
	4	5	7					
6				4	8	5		
							8	6
3			4		1			9
7	5							
		2	3	5				1
					7	6	9	
					2		3	

Hard # 818

	6	3		2			1	
		4	1					
8					6	4	5	
	9							7
	3						4	
7							2	
	5	6	4					9
					9	8		
	8			5		2	7	

Hard # 819

	2			9		5		
		9		7			1	
		1			4		9	7
9			7					
		6				8		
					5			4
6	4		5			1		
	1			3		6		
		2		6			8	

Hard # 820

	8		2			1		
5		9						
6			3	7				
3	7			5			6	
		5				9		
	1			2			5	3
				1	8			5
						3		9
		6			7		4	

Hard # 821

		3		2			4	
	1			6				
			9		8	1		
		6		3				
2		4	5		6	3		8
				9		2		
		1	4		9			
				7			3	
	8			5		7		

Hard # 822

							8	
	6	2	3		8	9		
		5	1	7				
4	8					7		
				6				
		9					6	4
				3	6	2		
		7	2		9	8	3	
	3							

Hard # 823

			2	9				
			4				2	8
5		8		7				
		3					8	1
	1		3		2		9	
8	7					3		
				5		6		4
7	4				9			
				4	3			

Hard # 824

			8				2	
8			5			6		1
				7				8
	4			3	7		6	
		2				7		
	1		4	2			8	
6				5				
9		1			6			5
	2				8			

Hard # 825

							8	
		3	2	4				1
			8	1		5	4	
	7						6	3
		6				8		
4	9						1	
	3	4		2	1			
8				9	6	1		
	2							

Hard # 826

						4		
	5	6		4	2			8
					7		5	
	2		1				9	
6		4				8		3
	8				6		2	
	9		4					
4			9	1		5	8	
		3						

Hard # 827

2		4				9		
				3	6			1
		9	1					
	5			2		3	7	
			8		5			
	7	2		9			5	
					3	7		
7			2	6				
		3				2		6

Hard # 828

4				7	5		9	1
6			3			8		
					1			
	4	7						9
	9						8	
1						7	5	
			1					
		2			7			3
3	6		9	2				4

Hard # 829

	7		1				6	
	4		9				2	
	6	1		5		9		
6	3							
			8		4			
							5	2
		3		4		6	8	
	2				3		1	
	5				8		4	

Hard # 830

6			9				7	
	1				6			
		7				2	4	
		8	4					3
	7			1			9	
1					3	4		
	6	1				9		
			2				1	
	3				8			5

Hard # 831

	8		9			7	6	2
4						1		
		9			8			
	5				1			
			3	8	9			
			2				3	
			1			8		
		2						7
7	9	4			6		2	

Hard # 832

7							5	
		8		1	9			6
		1				3		8
6			8		2			
	4						7	
			7		3			5
1		4				5		
2			5	9		6		
	6							3

Hard # 833

					9	4		
5	7				8			
2			6					
	6			7			1	5
	3						4	
9	1			8			3	
					3			1
			1				2	9
		8	9					

Hard # 834

				6	5	3		
9	5		3					
		1		4				
		5		2			3	
	2	3				6	7	
	1			9		5		
				5		7		
					6		4	8
		6	9	7				

Hard # 835

				3	5			
		4	7		2	5		
		2		1				4
		6			7			8
	7						5	
8			6			9		
9				4		3		
		1	9		3	4		
			2	8				

Hard # 836

		1	7	9	4			
	9				8			
	7			2			5	
8			2			3	1	
	6	7			3			9
	5			7			4	
			3				9	
			6	8	1	2		

Hard # 837

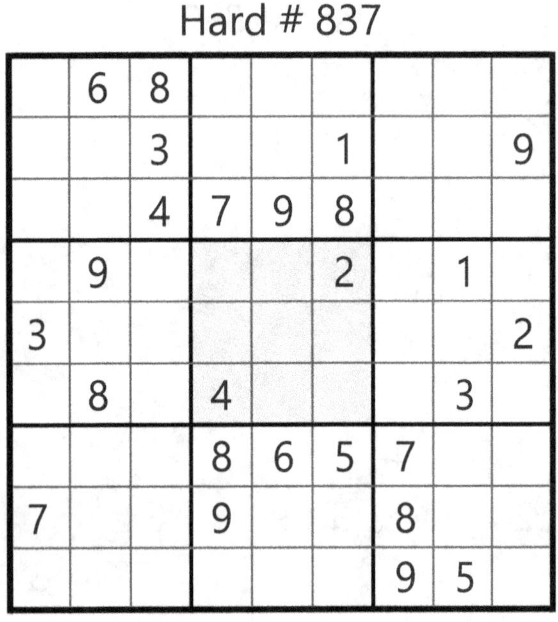

	6	8						
		3			1			9
		4	7	9	8			
	9				2		1	
3								2
	8		4				3	
			8	6	5	7		
7			9			8		
						9	5	

Hard # 838

9					3		1	8
			9	4		5		
							6	
		9		7		8		2
			5		8			
3		5		2		6		
	9							
		7		1	4			
2	1		8					7

Hard # 839

	2				5	1		7
8				2				
					9		2	
	7			9	6			8
5								6
4			8	5			9	
	5		1					
				4				3
6		4	5				7	

Hard # 840

	4		2				5	
6	5	3	8					
2				3				1
	8							
5		9				4		3
							9	
4				8				7
					7	6	2	9
	6				2		3	

Hard # 841

		3		6	5			
5								3
					8		4	
9		2				1	7	
	5						3	
	1	6				4		2
	2		1					
7								5
			4	2		7		

Hard # 842

2		5		9			8	6
4				5	8			1
		3						
			5			1		
	3						4	
		4			7			
						2		
6			4	2				9
5	9			7		3		4

Hard # 843

	8							6
7		2		8	1			
3						5		
				5	8	4		7
				3				
5		8	7	1				
		3						9
			9	7		1		8
1							3	

Hard # 844

	7			5	3		9	
				6		1		4
		6		9				
2					9			
	9	3				8	5	
			2					7
				7		5		
6		1		2				
	2		3	8			4	

Hard # 845

	1		3		4			8
7			2					
			8	5			4	
1	5					9		
6								5
		9					3	1
	6			4	3			
					2			3
8			5		7		2	

Hard # 846

	4		2					7
5	8	6		9				
	1		3	8		5		
		8		1		4		
		4		5	9		6	
				7		9	3	8
3					8		4	

Hard # 847

1			7					
				9	6		1	
	9	8		3				
7			4			3	8	
		5				6		
	2	3			1			7
				7		4	2	
	7		6	1				
					5			3

Hard # 848

			7				2	6
9	8		2	1	3			
2								
				6	2	5		
		2				9		
		4	1	3				
								5
			3	5	9		7	4
5	3				7			

Hard # 849

					3			1
	9							5
8		7			9			
		9		4		1		2
1				7				4
3		2		5		7		
			7			8		3
9							2	
4			6					

Hard # 850

				3	9		8	
		6				2		1
	7							
9		8		1		6		
1			3		8			2
		3		5		8		4
							5	
6		7				1		
	5		4	8				

Hard # 851

		2	6	8				
7		8			5		9	
		3			2			
							7	4
4				1				2
2	8							
			4			8		
	5		2			6		1
				9	8	3		

Hard # 852

	7			6		5		
	9	5						
3				8	5			
			9		8			6
	3			7			8	
1			4		2			
			3	9				8
						1	4	
		1		5			2	

Hard # 853

2			5		4			
7	8		1					5
		5		2				
			6			8	7	
	4						2	
	9	1			7			
				5		1		
9					1		8	2
			8		3			6

Hard # 854

3							2	
				4	7		1	
9		1		2	3			
	8	5			4			1
4			5			9	7	
			3	6		1		5
	4		7	9				
	2							7

Hard # 855

5				9			4	6
							1	
			4			7	3	
8						1	9	
			8	6	1			
	2	3						7
	3	5			6			
	1							
9	7			2				4

Hard # 856

9		1			3			6
	2	6			5			
				1			5	
6	4			3				
		2				7		
				6			8	9
	6			9				
			7			4	1	
8			3			6		2

Hard # 857

					1			9
5				4			7	
7				9	2	4		
	7	3						
2	8						9	4
						5	2	
		8	4	6				2
	6			8				7
9			1					

Hard # 858

		2		6				1
4	1							3
			8	9				
	8		9					6
	3						1	
5					8		7	
				4	2			
9							5	4
7				3		2		

Hard # 859

	9			2			6	
		2			7		3	
			6		9			1
8	5			4				3
7				5			1	8
3			2		1			
	8		5			1		
	2			6			9	

Hard # 860

		3	9				6	
	7	9			6			
8				2				
5					7	4		
	1	4				5	2	
		8	2					1
				3				9
			1			7	5	
	9				5	3		

Hard # 861

2	6				3			4
					6	8		5
		1						
8		3		7	2			
		9				6		
			4	8		3		2
						9		
6		5	7					
7			9				4	6

Hard # 862

		1				7	9	
			9	2				5
		9	1		7			8
	7				3	5		
		6	4				3	
3			7		9	2		
6				4	1			
	8	7				6		

Hard # 863

4					5			
7							2	
	3			1			6	8
	2		6	8				
		7	5		1	9		
				7	3		8	
5	1			3			4	
	6							7
			8					2

Hard # 864

6		1			9			7
5		7		4				
				3		2		
9	7					6	5	
	5	6					1	9
		2		7				
				2		3		6
7			8			1		2

Hard # 865

6	9			7	2	4		
			5					
4	1						2	
			1					9
8		1				7		6
3					6			
	3						8	5
					3			
		8	4	5			6	3

Hard # 866

1		6			2			
		2		7			8	
7					3	6		
		7	9		8		5	
	2		7		5	4		
		5	2					8
	9			6		7		
			5			9		1

Hard # 867

			6					
				2			7	3
	7	9		4			8	1
	6				4			
2	8						6	7
			1				5	
7	3			8		9	4	
8	4			9				
					6			

Hard # 868

1	7			4		8	6	
		6						
			3	8				
	5	1	4					
8	4						1	3
					7	9	4	
				9	3			
						2		
	3	4		5			7	9

Hard # 869

	4		2					6
			5				9	
		2		7	1			
		6				8	1	
9		7				5		2
	5	1				3		
			6	9		7		
	1				4			
7					5		8	

Hard # 870

		1	2				5	6
3				1				
		5			7			1
	4	3		2				
	5						1	
				3		5	6	
9			7			6		
				8				7
2	7				4	8		

Hard # 871

		7			4		1	2
				3	8			
5		4						
	7		3	5		8		
		5				6		
		2		4	7		3	
						2		1
			4	6				
7	5		8			9		

Hard # 872

					8	6		
9			4					7
		3		2				4
			9	5	6			
	8	1				5	9	
			8	1	4			
7				9		8		
5					2			1
		6	3					

Hard # 873

	7		6		1		2	
	1				9	5		
		2						7
	3		9			6		
			5		6			
		1			8		3	
7						1		
		4	2				8	
	2		1		3		5	

Hard # 874

8					3			
		5	7					
	3			1			8	5
				9		4		3
	4		6		7		9	
5		9		4				
1	6			5			3	
					2	9		
			1					8

Hard # 875

		9				6		2
	8		2			9		5
2					4			
7		4		8				
			9		7			
				5		8		3
			7					6
9		2			6		5	
1		3				2		

Hard # 876

4					3			8
		8				7		
	5				6	4		
	9		5				6	
1				8				4
	3				2		9	
		5	2				4	
		3				2		
2			1					7

Hard # 877

2	3				9			
9				4				
	7		2		8		4	
8						9		
			1	5	4			
		7						2
	1		7		3		2	
				1				7
			8				5	1

Hard # 878

					6		8	
8				9		3		
3			4				9	1
		7						9
			6	1	3			
4						5		
7	6				5			3
		5		3				4
	8		7					

Hard # 879

		4	1					
	8	6		2			9	
		7	3		6			
	3				7			6
	5						3	
8			4				2	
			9		1	3		
	6			3		2	4	
					2	9		

Hard # 880

	4	5					9	
				3	1		4	
		2			9			7
		9			7			
4	7						6	1
			8			2		
1			9			5		
	5		6	2				
	2					4	8	

Hard # 881

	2					7	1	
				8			5	6
9		6		3			8	
			4					
1	3						2	7
					6			
	9			5		8		2
7	6			4				
	5	2					6	

Hard # 882

8			7		2			
	1						9	
5	2				9	1		
	5		3	8				
1								5
				9	5		4	
		4	5				3	6
	6						8	
			6		8			4

Hard # 883

	6	4		8				
					9			
	5			3			7	2
	9	8			1			3
7								9
1			5			7	8	
2	8			1			6	
			3					
				6		5	9	

Hard # 884

1						3		
	4	6		1				
7		5		8		2		
					1		9	7
		7				5		
5	2		6					
		9		7		6		5
				6		1	3	
		1						8

Hard # 885

		3			8			
5		1	2					
	2	8						4
	5	2		9	1			
	4						6	
			4	6		2	5	
9						6	4	
					4	8		2
			8			9		

Hard # 886

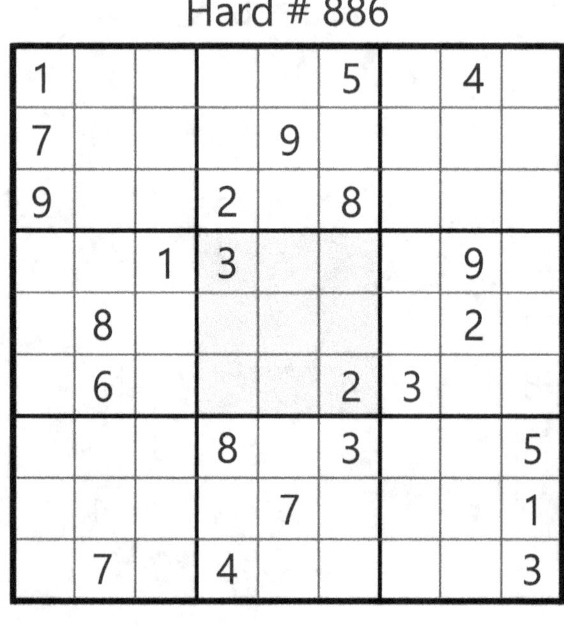

1					5		4	
7				9				
9			2		8			
		1	3				9	
	8						2	
	6				2	3		
			8		3			5
				7				1
	7		4					3

Hard # 887

	7				6			
2	1					6		
	6	5		8				2
3				7	4			
		9				7		
			3	5				9
5				2		1	3	
		4					6	7
			4				2	

Hard # 888

	7		5	2		8		
			3	4				7
					7			4
	5	4					9	
	9					5	6	
8			7					
1				8	4			
		3		1	5		7	

Hard # 889

		1				7		8
				4				
			8	3				2
	3					5	9	
4	8						2	6
	6	9					4	
8				6	5			
				9				
7		4				1		

Hard # 890

			5				6	8
				8			9	1
3			4					
	1				6	7	8	
		3				9		
	8	4	1				5	
					5			2
8	4			1				
2	6				4			

Hard # 891

		9		4			6	
		2						
		3	8	7			9	
	8				5			2
	7		3		9		1	
1			6				8	
	9			6	8	4		
						8		
	4			5		3		

Hard # 892

	3					4		
			3		5			1
					7		6	8
	4			9			1	6
			7		3			
2	7			1			4	
9	2		6					
5			1		9			
		4					3	

Hard # 893

			1			9		
	2			9		4		
5					6			
	8		6	5	3		1	
	5						8	
	7		8	1	9		3	
			9					2
		4		6			5	
		5			7			

Hard # 894

					6	9		
1	2		9					
		9		2		7		
	1			8	4			
	8	6				5	4	
			6	1			7	
		8		4		3		
					9		6	8
		5	2					

Hard # 895

3				4		1	7	
4						2		
1				7	2			4
6						8		
			2		8			
		5						7
7			9	6				1
		3						8
	1	9		5				3

Hard # 896

	1		5					
2					1	7	3	
				7	6			4
					3			8
		9		5		1		
5			8					
4			6	2				
	8	3	7					6
					5		7	

Hard # 897

4			3				6	7
			7					
	8	2				5		
1	7				8		4	
			9		3			
	9		6				5	2
		5				9	1	
					2			
9	4				6			3

Hard # 898

	8					5	4	
			5				2	1
6				2	4			8
		3	7					2
7					2	4		
2			1	9				3
9	6				8			
	3	5					8	

Hard # 899

	7							4
			1	2		9		
				9			3	7
	2		7			5	1	
		5				7		
	3	1			8		6	
3	6			8				
		7		1	2			
1							5	

Hard # 900

					2			
5		3						
4		8			7			2
	6			5			8	
3	5			6			2	9
	9			3			7	
9			8			4		5
						7		8
			9					

Hard # 901

1				2				8
	8			5				
	4	7	9				5	
			2					9
	3						4	
9					4			
	9				5	8	7	
				8			9	
4				7				2

Hard # 902

7			5			3	8	
5		2	9					6
				2				
9					6		4	
	3		1					5
				5				
2					9	6		8
	8	1			7			3

Hard # 903

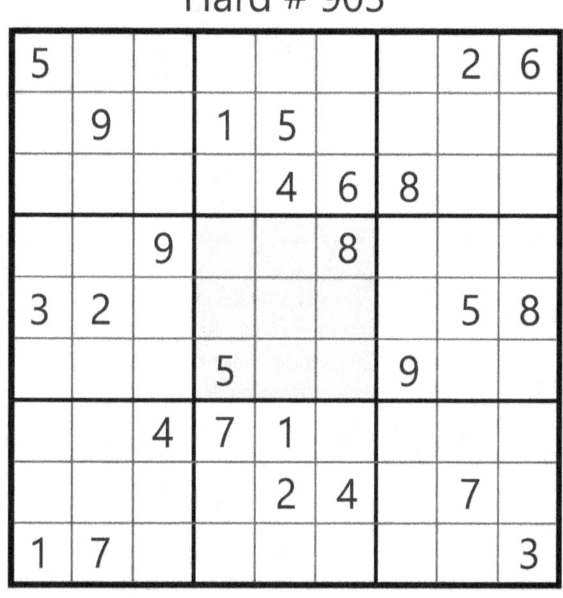

5							2	6
	9		1	5				
				4	6	8		
		9			8			
3	2						5	8
			5			9		
		4	7	1				
				2	4		7	
1	7							3

Hard # 904

			4		9	2		
		3		1	5			8
					3			1
	8							
	5	7	3		2	1	4	
							9	
4			7					
1			5	9		6		
		5	2		1			

Hard # 905

4			6			9		
	9		4	3			7	8
						4		
		5			2		9	
	3						2	
	6		9			5		
		1						
7	2			5	9		6	
		3			8			1

Hard # 906

4			3					9
	3	5					8	
				6	1	2		
		9	5		3			8
1			8		2	9		
		4	2	8				
	8					7	2	
7					5			6

Hard # 907

					2		7	
7		5				4		
1	6			3				
					3	8		2
		8				9		
4		3	1					
				8			5	9
		1				2		4
	3		5					

Hard # 908

6			4					
	1				8	5		
	3			1		8		9
			2					5
1			9		5			2
4					7			
8		9		7			5	
		7	8				9	
					3			8

Hard # 909

	1				7			
		9	6					
			4	9		5		7
		4	5			6	2	
			2		4			
	6	8			9	4		
9		2		4	6			
					1	3		
			9				7	

Hard # 910

				3		8		6
			4	8		7		
					2		3	
1			2			9		
5	9						1	7
		7			6			4
	3		1					
		6		2	4			
9		8		5				

Hard # 911

	1				7	3		
				5	4		7	
		6		2				
					9	8		6
			8	7	3			
8		5	2					
				8		5		
	8		7	3				
		2	4				1	

Hard # 912

					2		9	
	4	6		1	8			
						8		4
				4		1		2
		1				6		
5		8		7				
6		5						
			9	8		7	4	
	8		1					

Hard # 913

	2			7		8		
	5				6			
1					3			4
	9			4				6
8		5				1		7
4				1			3	
6			7					5
			2				7	
		3		9			1	

Hard # 914

	8				3			
2			6	4				
						6	8	
4				7			9	8
		7	9		6	1		
1	9			3				2
	5	1						
				8	9			7
			1				3	

Hard # 915

	1		8		6		9	
9	7			5		6		
					2			
3	2			8				
5								7
				1			6	9
			3					
		2		6			3	4
	9		7		1		8	

Hard # 916

7		6	4	3				
	5							
	9				8	3		
		5			2			6
		1	7		3	9		
3			8			4		
		4	6				8	
							1	
				7	1	6		4

Hard # 917

					7			2
	6	4	3					
		2	9	8		1		
	9			5				8
	4						1	
6				7			4	
		1		3	8	5		
					4	8	3	
9			7					

Hard # 918

			5		6			
						3	7	2
				7		1		
	2	3	7			6		
6	4						3	7
		7			9	2	1	
		2		3				
4	8	5						
			9		7			

Hard # 919

					5	3	8	
				2				4
	7	1					6	
9				8	6			
6	2						1	9
			2	3				6
	8					2	7	
1				7				
	6	7	3					

Hard # 920

				8	4	7		
1					6			
		6	3				8	
7			5			6		1
	3						9	
9		1			8			3
	9				1	4		
			8					2
		5	6	2				

Hard # 921

7	4				1	9		
8	5		9					
9			8		7			2
	6			2			1	
4			6		3			7
					8		7	6
		6	5				3	1

Hard # 922

				1	5	7	2	9
	6	7						
							3	4
3	5			8				
			9		6			
				2			4	8
4	7							
						4	7	
9	3	2	4	5				

Hard # 923

		6	8		9			
								7
	1					5		3
				4		9	7	1
	4		1		7		3	
1	6	7		8				
5		1					4	
9								
			2		8	3		

Hard # 924

2					7			
		5	1			9		
9	6		4	3				
4		3						6
	5						1	
6						5		4
				2	4		9	8
		6			5	4		
			8					7

Hard # 925

6			8	7				
9					2			1
		2		3		7		
	5				8		2	7
8	3		4				9	
		1		9		8		
2			3					4
				4	5			2

Hard # 926

		2	6			1		
		5						2
8			4				5	3
			9	7				
1	5						6	9
				1	8			
4	9				7			6
2						5		
		1			9	4		

Hard # 927

2	5							8
				5	1			6
		6	7					
				2	4	3		
	9		6		7		8	
		1	8	9				
					2	5		
1			3	8				
9							2	7

Hard # 928

	3			7				
			8					1
		8	6		1	5		
2							1	3
	5		9		2		4	
1	4							2
		6	3		8	2		
9					5			
				9			6	

Hard # 929

		1		4			5	
						7	1	3
	2				1			
			1			8		
8			3		7			4
		9			8			
			6				3	
4	5	7						
	9			7		2		

Hard # 930

			2					6
8		5		7		4		1
		6						
			8		1			5
5								2
3			9		5			
						1		
1		2		3		6		4
7					4			

Hard # 931

			5			9	6	
1			2			5		
		9	7		6			
	3		6				9	
				4				
	8				5		7	
			3		7	8		
		6			2			5
	7	2			1			

Hard # 932

9				6				
2			5					
7		8	2				5	
	9		7			1		3
1								8
6		3			4		7	
	7				3	8		5
					6			7
				5				4

Hard # 933

		7					5	3
		6					8	
	3		1	9				2
			5		3		6	
				4				
	5		7		6			
5				3	8		9	
	4					8		
2	7					1		

Hard # 934

3	5		4				1	8
			3					
8				7				2
5		4					6	
		3				2		
	6					4		9
2				6				1
					1			
6	1				5		9	3

Hard # 935

	6		2	7				
8					4			
	1						3	9
					1	7		
5				3				4
		9	6					
7	8						6	
			4					2
				9	5		1	

Hard # 936

9						8		1
			5				9	
		8	9		1			
1			2	3				
	5		8		9		7	
				1	7			5
			1		3	9		
	7				2			
4		2						3

Hard # 937

						1		
			9			8	4	
	1			7	5			
	6	3	1		9			4
7			5		4	2	9	
			8	6			3	
	3	5			2			
		4						

Hard # 938

		1	3					9
		7		9	6		8	
		8		2				
4						8	1	
		9				7		
	2	6						5
				3		5		
	3		4	8		6		
8					2	1		

Hard # 939

	7				8			4
					6	2		
			9	2			5	
		4				3		6
7	8						9	5
1		3				4		
	4			8	9			
		8	1					
9			5				7	

Hard # 940

	7			3	2	1		
4					9		5	8
	5	1	8			4		
8								3
		2			1	8	7	
1	2		6					9
		3	2	1			8	

Hard # 941

				4				
2		5	1		6			
	4		8			2	6	1
		9					2	
		6				7		
	8					9		
7	6	1			3		8	
			7		5	6		9
				1				

Hard # 942

6	1				4	7		
	2	9						
					1	9	8	
3					7			
			5	6	8			
			4					6
	5	3	8					
						3	2	
		2	3				6	7

Hard # 943

3					7		2	
5		7		4				1
	1	8				4		
			3					
			2	7	8			
					5			
		3				5	1	
6				2		3		7
	2		1					4

Hard # 944

			6	4	5		9	
		2						8
			7			4	3	
1		8						
	5	9				8	7	
						3		1
	6	5			3			
8						2		
	2		5	7	8			

Hard # 945

	9	7			6			
					2			7
			9				8	6
2							3	9
	6		4		3		7	
9	3							5
3	8				7			
5			1					
			8			9	5	

Hard # 946

		6	2	7				
7					5			
4		9			8		5	
				1	7	4		
		8				2		
		7	3	9				
	8		6			9		1
			9					8
				8	1	6		

Hard # 947

3		4				6		
		6	1					3
	1		2				5	
	3	9		5				
	5						8	
				9		5	1	
	7				4		2	
2					1	7		
		3				1		6

Hard # 948

	8		3				7	
		9		7			4	
4	5			6				8
	9							
5		6				8		7
							1	
7				2			5	3
	3			5		9		
	2				4		6	

Hard # 949

							7	4
2				3				5
	6				5		8	
		4			7	2		
			5		1			
		5	8			1		
	8		1				2	
9				2				3
7	3							

Hard # 950

4		6				1		
	7			8				9
			2		5			
	4	7						
8			7		9			5
						2	3	
			8		4			
9				3			5	
		5				8		1

Hard # 951

			2	5		4		
		3		7			9	
7			8				1	5
							3	2
		2				7		
4	9							
1	6				3			7
	3			2		1		
		9		6	8			

Hard # 952

4							3	
		3	2				1	4
	9				7	2	6	
				6				
5		6				4		3
				8				
	8	1	4				5	
2	3				9	8		
	4							1

Hard # 953

9	5	2	3					
1			4				3	
			5		7			
	1	8						
3		5				7		2
						9	1	
			8		9			
	2				4			9
					2	4	6	1

Hard # 954

			4	6			1	
6		7			3			
							7	3
		4	9					1
1			7		8			2
9					1	8		
2	3							
			8			3		4
	5			2	6			

Hard # 955

			5	2	3		7	
7		6				9		
				3		4	5	
		9	6	7	8	3		
	3	1		4				
		3				7		1
	8		3	9	4			

Hard # 956

	3			1		5		
	2		9					1
8		6						3
		3			2			
			1	4	3			
			5			8		
3						6		8
5					8		2	
		4		9			5	

Hard # 957

	8	3			5		9	
9				3		7		
	7			2		8	5	
					8			
1								9
			7					
	5	9		6			2	
		6		8				5
	3		5			9	1	

Hard # 958

			4			2		
				7		9	3	
			3	8	6			5
		2					5	1
		4				7		
3	9					6		
6			7	1	8			
	7	5		9				
		8			4			

Hard # 959

4							6	
			1		5		9	
	3			9		2	7	
					2		8	
		8		6		9		
	9		7					
	5	3		4			2	
	6		3		1			
	7							5

Hard # 960

			2			9	8	
		7		5	9		2	4
					7			5
2		5		1				
				8		5		3
8			4					
3	5		7	2		4		
	7	6			3			

Hard # 961

4			3				5	
	9	1	4					
	7			5		4		
2					3			
7		6				5		4
			7					1
		2		8			6	
					6	3	8	
	6				9			5

Hard # 962

								6
3		9			6		1	5
		6	8	5				
			9	1				2
		8				1		
7				8	3			
				2	7	5		
2	3		4			7		1
1								

Hard # 963

		5			6		3	
	2			4				1
9			7			4		
3	9		4			2		
		2			1		7	3
		4			9			2
7				1			6	
	3		5			8		

Hard # 964

					4		6	2
						3		
			5	3	7			1
	7				9	6		
	4			8			9	
		1	3				7	
4			9	1	6			
		8						
7	6		4					

Hard # 965

					6		2	
		2				6		
6	8				9			
7				2		3		
	1	8	3		4	9	6	
		9		1				5
			1				8	6
		7				1		
	3		5					

Hard # 966

					5			8
	5		6					
		7		9		5		
1				8		7		4
		6	2		1	9		
3		9		4				1
		4		7		1		
					9		5	
9			3					

Hard # 967

5					7			2
							6	
	7		1	4				
		4	7	1		5		
		3	4		2	6		
		7		9	6	1		
				2	5		8	
	6							
7			9					4

Hard # 968

	7			3				1
						2	6	
2				5	8			
		2		7			5	
		8	3		9	6		
	9			8		3		
			8	4				6
	5	7						
9				1			4	

Hard # 969

	4	3	6			9		
						4	1	
				1			8	
	2	4			6	8		
			8		7			
		1	4			7	5	
	1			9				
	3	5						
		8			3	1	9	

Hard # 970

2		6	3					7
		3	7					
		8		9				
3					9			1
		7	6		1	3		
4			5					6
				8		2		
					2	6		
8					3	9		5

Hard # 971

	6				7		3	
2	7	9			4			
				6				
1					8	3		
3	2						1	9
		4	2					7
				8				
			9			6	4	1
	3		5				7	

Hard # 972

		3			4	1	9	
						3		6
				7	1	2	8	
7							2	
			4		6			
	6							1
	2	9	5	3				
5		7						
	4	1	7			9		

Hard # 973

	8				9			5
		1		6			2	
	2		1					
5		6	8					
		2		4		8		
					5	4		3
					2		7	
	3			7		9		
7			9				1	

Hard # 974

				9	6			4
4				2			3	
	2		4					6
	1	5	7			9		
		8			9	5	4	
7					2		9	
	5			7				8
9			5	6				

Hard # 975

		9			8			
					2	5		
	4	1		7				6
	3	5			1			
1								5
			6			2	9	
8				2		9	1	
		2	5					
			7			4		

Hard # 976

5				3				
	3	6	1					8
		2	7			6		
	2				1			
4		1				3		9
			4				1	
		8			4	2		
6					7	1	4	
				9				6

Hard # 977

4	3			2	1			
		6		7	8	5		4
3	1	8						
			9		4			
						6	3	7
6		3	8	4		9		
			5	6			8	3

Hard # 978

1			8			4		
							7	
	2	3			5	1		
				9				6
		1	2	4	3	7		
3				8				
		4	7			6	2	
	8							
		2			9			8

Hard # 979

				6	7		2	8
6			1			4		
8					4			
		5					3	2
7								5
3	1					8		
			4					7
		4			5			9
9	5		2	3				

Hard # 980

9	5	7		3	4			
							8	
3				7				
2	7		3			9		
		4				3		
		3			6		1	7
				5				6
	2							
			1	6		8	3	5

Hard # 981

			5				7	1
5	3			7				
	4			3	6	5		
						9		2
			3		4			
1		6						
		4	1	2			8	
				6			2	9
6	1				5			

Hard # 982

	3		1					6
	8	9						3
6	4					9		
9			2	5			4	
	1			8	4			7
		4					3	2
3						6	7	
1					7		9	

Hard # 983

4						5		
		7		1				
	2				8		6	4
		9		5		8	7	
	5	6		2		9		
3	8		9				4	
				6		7		
		2						3

Hard # 984

	7	6	4					3
				9			6	
			6	8				4
		3		4				
5	4						9	2
				3		1		
2				5	8			
	6			2				
3					7	2	5	

Hard # 985

	6		1		8			
	2	5						6
	7	1		3				
				1				3
		4	6		7	9		
7				8				
				6		5	4	
9						8	6	
			9		5		1	

Hard # 986

4		1						
		9	2			6		
				8	1	2		
	3		7				5	
		2		4		9		
	9				8		2	
		6	1	7				
		7			5	3		
						8		1

Hard # 987

	1	4					8	
	6		1					9
				5		4		
	7					1		5
		1	3		9	7		
8		2					6	
		6		8				
1					2		5	
	2					6	7	

Hard # 988

		1		2		8		
			5	3				
5	7				9			1
7								8
	9	4				2	1	
8								6
6			1				7	9
				9	2			
		9		7		3		

Hard # 989

			7	9	2		6	8
		6		4				
	8						4	
9					5	1		
5								2
		1	9					3
	4						2	
				5		8		
7	2		6	8	1			

Hard # 990

		2		9				
		9	3					2
	3	1				4		8
5			6		1		8	
	9		8		2			3
2		7				8	3	
9					4	1		
				6		2		

Hard # 991

		5						3
9	4							
		2		9	8		1	
5		9	4	7				
	2						9	
				3	1	2		5
	8		7	1		5		
							4	7
4						8		

Hard # 992

	4			3	8			
							9	
	8	5	9		2	1		
	2			9				7
			3		7			
3				2			6	
		3	6		1	8	2	
	1							
			7	4			5	

Hard # 993

					5			
2	4		6				1	
	3	9	2					
4			3			8		
3		8				7		4
		1			2			9
					8	1	7	
	2				7		8	3
			1					

Hard # 994

		1		5			6	
		9				3		4
	6				2			
8							7	
	7	3	5		1	4	9	
	2							1
			4				1	
3		6				9		
	8			6		2		

Hard # 995

				7				
		9			1			
	5	3	4			2	1	
3	1					6		8
	4						7	
6		8					4	3
	7	2			3	1	5	
			8			7		
				2				

Hard # 996

		6		9	7		2	4
	4							1
3								6
					2		8	
6			3		5			7
	1		8					
1								9
8							1	
4	3		6	2		7		

Hard # 997

			1					5
				8	5			2
			3				8	6
	4		9		3		6	
	3						2	
	6		4		8		3	
5	1				4			
2			8	9				
9					1			

Hard # 998

2		1						
	7		3	1	6			
			4					5
		4	5	6			1	
	9						5	
	8			9	7	4		
7					2			
			9	3	1		8	
						2		6

Hard # 999

	4		5				3	
						9		1
3	8			2				
8			2		3			
		9				8		
			4		1			5
				6			1	4
5		8						
	6				7		2	

Hard # 1000

8								4
		4	9					5
5		1		6				7
			4			2		
		8		5		9		
		7			6			
3				7		4		9
1					9	6		
2								3

Hard # 1001

	8					3		
4	7				8		2	
		6	4	2				
	4				7	8		
6								7
		5	2				1	
				3	9	1		
	1		8				5	3
		7					8	

Hard # 1002

	3					1		
6			8	9		3	7	
8	7							
		4		2				3
			4		9			
7				3		5		
							3	4
	2	3		5	6			9
		6					5	

Hard # 1003

	3					4		5
								2
			9	5	1		3	
	5	6		8		2		
				6				
		3		9		6	4	
	7		2	4	8			
1								
6		8					9	

Hard # 1004

	1	2		5	6		7	
		5						
		9	2		4			
	5		1					
7			4		8			1
					3		8	
			6		1	5		
						2		
	6		3	7		8	1	

Hard # 1005

	7				5	8		
		1		7		9		
		9					6	3
		7	5				8	
	8						1	
	3				8	2		
2	1					6		
		8		2		5		
		6	4				3	

Hard # 1006

	4				7			
1			9			3	6	
		5		8				
	3		2				9	1
8								4
2	6				1		8	
				6		1		
	1	9			2			7
			7				5	

Hard # 1007

							4	
		7	8					9
	3	8		6		5		
				5			6	2
1		2				8		5
5	7			2				
		4		8		2	7	
9					4	1		
	8							

Hard # 1008

	2							
	3		8			1	7	
	9				7		8	2
3			7		8			
			1		2			
			9		5			3
2	4		6				1	
	7	9			3		6	
							2	

Solution

Solution # 1

7	3	5	4	1	8	2	9	6
9	8	6	5	7	2	3	4	1
1	4	2	6	3	9	5	7	8
3	7	9	1	6	5	4	8	2
2	5	8	9	4	3	6	1	7
4	6	1	8	2	7	9	3	5
5	2	3	7	9	1	8	6	4
8	1	4	3	5	6	7	2	9
6	9	7	2	8	4	1	5	3

Solution # 2

2	1	8	6	4	9	3	7	5
9	6	3	7	5	1	4	2	8
7	5	4	3	8	2	9	1	6
4	7	2	9	6	3	8	5	1
8	3	1	4	7	5	6	9	2
6	9	5	2	1	8	7	4	3
1	8	9	5	3	4	2	6	7
5	4	7	8	2	6	1	3	9
3	2	6	1	9	7	5	8	4

Solution # 3

6	5	3	8	9	4	2	1	7
9	7	2	3	5	1	8	4	6
4	8	1	2	7	6	9	3	5
5	1	7	9	4	8	3	6	2
3	4	9	7	6	2	1	5	8
8	2	6	5	1	3	4	7	9
1	6	8	4	2	5	7	9	3
2	9	5	1	3	7	6	8	4
7	3	4	6	8	9	5	2	1

Solution # 4

4	3	5	2	6	1	7	8	9
2	6	9	7	8	4	5	1	3
1	7	8	3	5	9	6	4	2
7	9	3	5	2	8	4	6	1
8	1	2	6	4	7	9	3	5
6	5	4	9	1	3	2	7	8
9	8	6	4	3	5	1	2	7
5	2	1	8	7	6	3	9	4
3	4	7	1	9	2	8	5	6

Solution # 5

7	3	9	4	5	1	8	6	2
2	5	1	8	6	7	4	3	9
8	4	6	3	2	9	7	1	5
6	7	5	1	3	4	2	9	8
4	2	3	9	8	6	1	5	7
9	1	8	5	7	2	6	4	3
1	6	2	7	9	5	3	8	4
3	9	7	6	4	8	5	2	1
5	8	4	2	1	3	9	7	6

Solution # 6

6	4	5	3	2	1	8	7	9
8	7	2	6	5	9	3	1	4
9	1	3	7	8	4	2	5	6
2	5	9	8	1	7	4	6	3
3	8	7	4	6	2	1	9	5
1	6	4	5	9	3	7	8	2
4	3	8	9	7	6	5	2	1
5	2	6	1	4	8	9	3	7
7	9	1	2	3	5	6	4	8

Solution # 7

4	3	1	7	9	5	6	2	8
7	9	2	8	3	6	1	5	4
8	6	5	2	1	4	3	7	9
3	4	6	1	7	8	5	9	2
5	1	9	3	4	2	8	6	7
2	8	7	6	5	9	4	3	1
9	7	4	5	8	3	2	1	6
6	5	8	9	2	1	7	4	3
1	2	3	4	6	7	9	8	5

Solution # 8

4	1	2	7	3	5	6	9	8
5	9	6	8	1	4	2	3	7
7	8	3	9	2	6	5	4	1
2	7	9	5	8	3	4	1	6
8	3	4	2	6	1	9	7	5
1	6	5	4	7	9	8	2	3
6	5	1	3	4	2	7	8	9
9	2	8	1	5	7	3	6	4
3	4	7	6	9	8	1	5	2

Solution # 9

7	3	9	2	5	1	4	6	8
8	2	6	4	9	7	1	3	5
5	1	4	3	6	8	9	2	7
1	9	8	6	3	2	5	7	4
6	4	2	7	1	5	8	9	3
3	5	7	8	4	9	2	1	6
4	7	1	5	2	6	3	8	9
9	6	5	1	8	3	7	4	2
2	8	3	9	7	4	6	5	1

Solution # 10

8	9	1	3	4	7	2	5	6
3	5	6	9	2	1	4	8	7
7	4	2	8	6	5	1	9	3
1	7	8	5	9	6	3	4	2
4	3	9	2	7	8	5	6	1
2	6	5	1	3	4	8	7	9
9	2	4	7	5	3	6	1	8
6	8	7	4	1	2	9	3	5
5	1	3	6	8	9	7	2	4

Solution # 11

4	7	6	3	1	8	2	5	9
8	5	9	6	4	2	7	3	1
2	3	1	7	9	5	8	6	4
6	8	7	2	5	4	1	9	3
3	4	2	1	6	9	5	8	7
9	1	5	8	3	7	4	2	6
7	2	3	9	8	1	6	4	5
5	9	8	4	7	6	3	1	2
1	6	4	5	2	3	9	7	8

Solution # 12

8	4	3	2	5	7	9	6	1
2	7	6	8	9	1	5	4	3
1	9	5	3	6	4	8	2	7
9	3	8	4	7	5	2	1	6
6	2	4	1	8	3	7	9	5
5	1	7	6	2	9	4	3	8
3	8	1	7	4	2	6	5	9
4	6	9	5	1	8	3	7	2
7	5	2	9	3	6	1	8	4

Solution # 13

4	6	2	9	7	1	3	5	8
3	9	8	6	5	2	1	4	7
7	1	5	3	4	8	6	9	2
5	2	9	4	8	6	7	3	1
1	3	6	5	2	7	9	8	4
8	4	7	1	9	3	2	6	5
9	8	1	2	6	5	4	7	3
2	7	4	8	3	9	5	1	6
6	5	3	7	1	4	8	2	9

Solution # 14

5	6	2	9	8	1	7	3	4
9	8	4	6	7	3	1	5	2
7	3	1	4	2	5	6	9	8
2	7	5	3	1	8	4	6	9
3	4	6	5	9	7	2	8	1
1	9	8	2	4	6	3	7	5
6	5	9	1	3	2	8	4	7
8	2	3	7	5	4	9	1	6
4	1	7	8	6	9	5	2	3

Solution # 15

9	4	5	7	1	6	3	8	2
8	7	3	4	5	2	9	6	1
6	2	1	9	8	3	4	5	7
5	6	7	2	3	4	8	1	9
1	8	2	5	6	9	7	3	4
3	9	4	8	7	1	6	2	5
4	3	6	1	9	5	2	7	8
7	5	9	3	2	8	1	4	6
2	1	8	6	4	7	5	9	3

Solution # 16

1	7	5	9	2	8	4	3	6
4	9	6	5	7	3	1	2	8
2	3	8	1	6	4	9	5	7
3	8	2	7	1	5	6	4	9
9	1	4	6	3	2	8	7	5
6	5	7	8	4	9	2	1	3
5	4	1	3	8	6	7	9	2
8	2	9	4	5	7	3	6	1
7	6	3	2	9	1	5	8	4

Solution # 17

2	4	7	8	6	1	9	3	5
5	3	8	2	7	9	6	1	4
6	9	1	4	5	3	2	7	8
4	6	3	5	2	8	1	9	7
8	1	9	7	3	4	5	2	6
7	5	2	9	1	6	4	8	3
9	2	4	6	8	7	3	5	1
3	7	5	1	4	2	8	6	9
1	8	6	3	9	5	7	4	2

Solution # 18

8	9	7	6	3	5	2	1	4
5	6	4	9	1	2	3	8	7
2	1	3	7	4	8	6	5	9
4	8	5	1	6	3	7	9	2
3	2	6	8	9	7	1	4	5
1	7	9	2	5	4	8	6	3
7	4	1	5	2	6	9	3	8
6	3	8	4	7	9	5	2	1
9	5	2	3	8	1	4	7	6

Solution # 19

6	9	5	3	4	8	7	2	1
1	2	3	9	7	5	8	4	6
4	7	8	2	1	6	5	9	3
3	8	9	6	2	1	4	7	5
2	6	7	5	9	4	3	1	8
5	4	1	8	3	7	2	6	9
7	5	4	1	8	9	6	3	2
8	1	2	4	6	3	9	5	7
9	3	6	7	5	2	1	8	4

Solution # 20

7	5	4	1	2	3	8	6	9
9	6	3	5	4	8	2	1	7
8	2	1	6	7	9	5	4	3
6	1	9	2	5	4	7	3	8
3	8	5	7	9	6	4	2	1
4	7	2	3	8	1	6	9	5
5	9	7	4	3	2	1	8	6
1	4	8	9	6	5	3	7	2
2	3	6	8	1	7	9	5	4

Solution # 21

1	6	9	4	7	3	2	8	5
4	8	7	2	1	5	3	6	9
5	2	3	6	8	9	7	1	4
7	1	6	9	5	4	8	3	2
9	3	4	8	2	1	5	7	6
2	5	8	3	6	7	4	9	1
6	9	5	7	3	2	1	4	8
8	7	2	1	4	6	9	5	3
3	4	1	5	9	8	6	2	7

Solution # 22

3	7	9	1	4	5	8	2	6
6	4	5	8	2	9	3	7	1
2	8	1	3	6	7	4	9	5
5	3	6	9	7	8	2	1	4
1	2	8	4	5	3	7	6	9
7	9	4	6	1	2	5	8	3
9	5	7	2	3	1	6	4	8
4	1	3	7	8	6	9	5	2
8	6	2	5	9	4	1	3	7

Solution # 23

8	1	9	5	4	7	6	2	3
3	5	7	9	6	2	8	1	4
4	2	6	1	3	8	7	5	9
5	6	1	2	9	3	4	7	8
2	8	3	4	7	1	9	6	5
9	7	4	8	5	6	2	3	1
7	4	5	6	1	9	3	8	2
6	9	8	3	2	5	1	4	7
1	3	2	7	8	4	5	9	6

Solution # 24

2	5	7	1	3	9	8	4	6
3	4	9	5	8	6	2	7	1
8	1	6	7	4	2	5	9	3
6	8	3	4	5	1	7	2	9
9	7	5	8	2	3	6	1	4
4	2	1	9	6	7	3	8	5
1	6	8	2	9	5	4	3	7
7	3	2	6	1	4	9	5	8
5	9	4	3	7	8	1	6	2

Solution # 25

2	7	8	4	6	3	1	9	5
5	1	4	7	9	2	6	3	8
6	9	3	1	5	8	2	4	7
7	4	9	8	2	5	3	1	6
3	6	1	9	4	7	8	5	2
8	5	2	6	3	1	4	7	9
9	3	5	2	1	6	7	8	4
1	2	7	5	8	4	9	6	3
4	8	6	3	7	9	5	2	1

Solution # 26

8	2	4	5	7	9	3	1	6
3	5	6	1	4	8	7	2	9
7	1	9	3	2	6	8	5	4
4	7	8	2	9	5	6	3	1
9	6	2	8	1	3	4	7	5
5	3	1	7	6	4	2	9	8
1	8	5	6	3	7	9	4	2
6	9	7	4	5	2	1	8	3
2	4	3	9	8	1	5	6	7

Solution # 27

1	4	2	3	8	6	9	7	5
8	3	7	9	1	5	2	6	4
5	6	9	2	4	7	8	1	3
2	8	1	4	5	3	7	9	6
7	9	3	1	6	8	4	5	2
6	5	4	7	9	2	1	3	8
9	1	6	5	2	4	3	8	7
3	2	8	6	7	1	5	4	9
4	7	5	8	3	9	6	2	1

Solution # 28

6	3	7	5	9	8	1	2	4
8	9	2	4	1	3	6	5	7
5	1	4	2	7	6	3	9	8
3	8	1	7	6	9	5	4	2
2	7	6	1	5	4	9	8	3
9	4	5	8	3	2	7	6	1
1	2	9	3	8	5	4	7	6
4	5	3	6	2	7	8	1	9
7	6	8	9	4	1	2	3	5

Solution # 29

4	9	2	5	8	1	7	6	3
7	1	5	3	4	6	2	8	9
6	8	3	9	2	7	1	4	5
9	6	7	8	1	2	3	5	4
5	2	4	6	7	3	9	1	8
8	3	1	4	5	9	6	7	2
2	5	9	7	6	4	8	3	1
3	7	8	1	9	5	4	2	6
1	4	6	2	3	8	5	9	7

Solution # 30

6	9	2	5	3	7	1	4	8
8	7	4	9	1	2	6	5	3
3	5	1	6	8	4	7	2	9
5	6	3	4	7	8	9	1	2
9	2	8	3	5	1	4	7	6
4	1	7	2	9	6	8	3	5
7	3	9	1	6	5	2	8	4
1	4	6	8	2	3	5	9	7
2	8	5	7	4	9	3	6	1

Solution # 31

5	8	2	3	6	4	1	7	9
1	4	3	7	2	9	5	8	6
7	6	9	1	5	8	4	2	3
4	5	6	2	1	7	9	3	8
8	3	7	9	4	6	2	1	5
9	2	1	8	3	5	7	6	4
3	9	5	6	7	2	8	4	1
6	7	8	4	9	1	3	5	2
2	1	4	5	8	3	6	9	7

Solution # 32

3	8	1	9	2	4	7	5	6
2	9	6	5	8	7	3	4	1
7	4	5	6	1	3	8	2	9
6	5	8	4	3	2	9	1	7
9	1	2	7	6	5	4	8	3
4	7	3	8	9	1	5	6	2
8	6	7	1	5	9	2	3	4
1	3	9	2	4	8	6	7	5
5	2	4	3	7	6	1	9	8

Solution # 33

8	4	3	6	7	5	2	1	9
7	5	9	2	3	1	4	6	8
2	6	1	9	8	4	5	3	7
5	3	4	8	6	9	7	2	1
1	8	6	7	5	2	3	9	4
9	7	2	1	4	3	8	5	6
3	1	5	4	9	7	6	8	2
4	9	8	3	2	6	1	7	5
6	2	7	5	1	8	9	4	3

Solution # 34

5	8	6	4	9	1	2	7	3
7	2	9	3	5	6	1	8	4
3	1	4	7	8	2	9	5	6
2	6	5	1	4	9	7	3	8
1	7	3	6	2	8	5	4	9
9	4	8	5	7	3	6	1	2
6	9	1	8	3	5	4	2	7
4	3	2	9	1	7	8	6	5
8	5	7	2	6	4	3	9	1

Solution # 35

9	7	5	8	6	2	4	3	1
4	3	2	1	5	9	7	6	8
6	8	1	7	3	4	9	2	5
2	5	9	4	7	1	3	8	6
8	1	4	3	2	6	5	9	7
3	6	7	5	9	8	1	4	2
7	9	3	6	8	5	2	1	4
1	2	8	9	4	7	6	5	3
5	4	6	2	1	3	8	7	9

Solution # 36

3	4	6	8	2	9	5	7	1
9	7	1	3	6	5	4	2	8
2	5	8	7	4	1	3	9	6
5	6	4	9	1	3	2	8	7
7	2	3	6	5	8	9	1	4
8	1	9	4	7	2	6	3	5
1	3	2	5	8	6	7	4	9
4	8	5	2	9	7	1	6	3
6	9	7	1	3	4	8	5	2

Solution # 37

2	7	4	9	6	3	8	1	5
9	8	5	1	7	4	3	2	6
6	1	3	5	8	2	4	9	7
5	3	9	6	4	8	2	7	1
1	4	7	3	2	9	6	5	8
8	6	2	7	5	1	9	3	4
3	5	8	2	1	6	7	4	9
7	2	6	4	9	5	1	8	3
4	9	1	8	3	7	5	6	2

Solution # 38

2	9	8	5	7	1	3	6	4
3	1	4	6	9	2	7	5	8
6	5	7	3	8	4	9	2	1
9	6	3	1	5	8	2	4	7
8	2	1	7	4	6	5	9	3
7	4	5	2	3	9	8	1	6
5	7	2	4	1	3	6	8	9
1	8	6	9	2	7	4	3	5
4	3	9	8	6	5	1	7	2

Solution # 39

7	6	8	2	9	4	3	5	1
1	9	2	5	7	3	8	4	6
4	5	3	8	1	6	7	9	2
2	3	7	6	4	9	5	1	8
8	1	5	3	2	7	4	6	9
9	4	6	1	5	8	2	3	7
3	7	1	9	8	5	6	2	4
5	8	9	4	6	2	1	7	3
6	2	4	7	3	1	9	8	5

Solution # 40

9	1	8	7	3	5	6	4	2
4	3	7	2	6	1	8	5	9
6	2	5	8	4	9	7	1	3
7	6	4	3	9	2	5	8	1
5	8	2	4	1	6	3	9	7
1	9	3	5	8	7	4	2	6
2	7	6	1	5	4	9	3	8
3	5	9	6	2	8	1	7	4
8	4	1	9	7	3	2	6	5

Solution # 41

7	9	3	4	5	8	1	6	2
6	4	1	7	2	9	3	5	8
8	5	2	3	6	1	9	7	4
4	3	8	1	9	5	7	2	6
5	6	9	2	8	7	4	3	1
2	1	7	6	3	4	5	8	9
3	7	6	9	1	2	8	4	5
9	8	4	5	7	6	2	1	3
1	2	5	8	4	3	6	9	7

Solution # 42

2	5	7	8	6	1	3	9	4
8	1	9	4	3	2	7	6	5
3	6	4	5	7	9	8	2	1
1	8	5	9	2	6	4	3	7
4	9	6	7	8	3	5	1	2
7	3	2	1	4	5	6	8	9
9	4	8	3	1	7	2	5	6
5	2	3	6	9	4	1	7	8
6	7	1	2	5	8	9	4	3

Solution # 43

2	7	3	9	5	1	6	4	8
8	9	6	3	4	2	7	5	1
5	1	4	8	6	7	9	3	2
7	5	9	2	1	6	3	8	4
1	3	2	7	8	4	5	6	9
6	4	8	5	9	3	1	2	7
3	6	7	4	2	9	8	1	5
4	8	1	6	7	5	2	9	3
9	2	5	1	3	8	4	7	6

Solution # 44

6	8	3	4	1	9	2	5	7
9	7	1	2	6	5	4	3	8
5	2	4	7	3	8	6	1	9
8	4	9	5	2	6	3	7	1
7	3	5	9	4	1	8	2	6
2	1	6	8	7	3	5	9	4
4	5	2	1	8	7	9	6	3
3	9	7	6	5	4	1	8	2
1	6	8	3	9	2	7	4	5

Solution # 45

4	8	3	1	6	5	9	7	2
7	9	2	4	8	3	6	1	5
5	6	1	2	9	7	3	4	8
2	3	9	8	7	6	4	5	1
6	1	4	5	3	2	8	9	7
8	5	7	9	4	1	2	3	6
3	2	8	7	1	4	5	6	9
1	4	5	6	2	9	7	8	3
9	7	6	3	5	8	1	2	4

Solution # 46

9	7	4	6	1	5	2	3	8
6	3	2	8	7	9	5	4	1
8	5	1	2	3	4	6	9	7
4	9	7	1	5	8	3	2	6
1	6	5	4	2	3	7	8	9
3	2	8	9	6	7	4	1	5
2	4	9	5	8	6	1	7	3
5	8	3	7	4	1	9	6	2
7	1	6	3	9	2	8	5	4

Solution # 47

7	4	9	2	8	6	1	5	3
5	8	2	1	3	4	6	7	9
1	6	3	5	7	9	2	8	4
6	9	4	8	1	3	5	2	7
8	2	5	9	6	7	4	3	1
3	1	7	4	5	2	8	9	6
2	7	8	3	4	1	9	6	5
9	3	1	6	2	5	7	4	8
4	5	6	7	9	8	3	1	2

Solution # 48

9	8	7	4	5	6	2	3	1
4	3	5	8	1	2	6	9	7
2	1	6	7	3	9	4	8	5
3	2	4	9	7	8	5	1	6
1	6	8	3	4	5	9	7	2
7	5	9	2	6	1	3	4	8
8	7	2	6	9	4	1	5	3
6	9	1	5	8	3	7	2	4
5	4	3	1	2	7	8	6	9

Solution # 49

2	4	3	9	6	1	7	8	5
1	7	9	8	2	5	6	4	3
8	6	5	7	3	4	9	2	1
9	5	7	6	4	8	1	3	2
6	2	1	5	9	3	4	7	8
4	3	8	1	7	2	5	9	6
7	8	2	4	1	6	3	5	9
5	9	6	3	8	7	2	1	4
3	1	4	2	5	9	8	6	7

Solution # 50

1	9	7	8	6	2	5	3	4
4	5	8	7	3	9	1	2	6
3	6	2	4	5	1	9	8	7
7	3	9	6	2	5	8	4	1
6	1	4	3	9	8	2	7	5
2	8	5	1	7	4	6	9	3
5	7	1	9	8	3	4	6	2
8	4	6	2	1	7	3	5	9
9	2	3	5	4	6	7	1	8

Solution # 51

2	4	5	9	1	3	7	6	8
7	8	1	4	2	6	5	3	9
9	3	6	5	7	8	4	2	1
3	5	2	1	4	9	6	8	7
1	6	7	8	3	5	2	9	4
4	9	8	2	6	7	3	1	5
8	7	4	6	9	2	1	5	3
6	1	9	3	5	4	8	7	2
5	2	3	7	8	1	9	4	6

Solution # 52

4	2	8	5	6	1	3	7	9
1	7	6	9	2	3	4	8	5
3	5	9	4	8	7	1	6	2
6	8	4	7	1	9	2	5	3
7	9	2	8	3	5	6	1	4
5	3	1	6	4	2	7	9	8
8	4	7	3	5	6	9	2	1
2	6	5	1	9	4	8	3	7
9	1	3	2	7	8	5	4	6

Solution # 53

6	5	8	1	4	7	2	3	9
3	9	1	8	2	6	5	7	4
7	4	2	5	3	9	8	1	6
1	2	5	3	9	4	6	8	7
4	3	7	6	8	2	1	9	5
8	6	9	7	5	1	3	4	2
2	1	4	9	6	8	7	5	3
5	7	6	4	1	3	9	2	8
9	8	3	2	7	5	4	6	1

Solution # 54

5	2	4	7	6	9	3	1	8
7	9	1	8	5	3	2	6	4
3	8	6	2	4	1	5	9	7
2	1	3	6	7	4	8	5	9
9	7	8	3	2	5	1	4	6
4	6	5	1	9	8	7	3	2
8	5	7	4	1	6	9	2	3
6	3	9	5	8	2	4	7	1
1	4	2	9	3	7	6	8	5

Solution # 55

6	3	9	5	7	1	2	4	8
4	1	5	8	3	2	7	9	6
8	7	2	4	6	9	1	3	5
7	9	3	1	5	8	6	2	4
5	2	4	6	9	7	3	8	1
1	6	8	2	4	3	9	5	7
2	8	6	3	1	4	5	7	9
3	5	7	9	8	6	4	1	2
9	4	1	7	2	5	8	6	3

Solution # 56

3	4	1	6	8	9	5	7	2
8	9	2	5	1	7	4	6	3
7	5	6	4	3	2	1	9	8
5	1	7	3	6	4	8	2	9
2	6	4	1	9	8	3	5	7
9	8	3	2	7	5	6	1	4
4	3	5	9	2	1	7	8	6
6	7	9	8	5	3	2	4	1
1	2	8	7	4	6	9	3	5

Solution # 57

4	8	2	3	6	5	9	7	1
1	3	5	9	2	7	6	8	4
9	7	6	8	1	4	2	5	3
6	1	8	4	9	3	7	2	5
7	2	9	6	5	1	4	3	8
5	4	3	2	7	8	1	6	9
8	9	7	1	3	6	5	4	2
2	5	4	7	8	9	3	1	6
3	6	1	5	4	2	8	9	7

Solution # 58

5	2	3	9	1	4	7	6	8
4	9	6	8	7	5	2	3	1
1	7	8	6	3	2	9	5	4
3	5	4	1	2	7	6	8	9
2	1	9	3	6	8	5	4	7
6	8	7	4	5	9	3	1	2
8	3	1	7	9	6	4	2	5
9	4	5	2	8	3	1	7	6
7	6	2	5	4	1	8	9	3

Solution # 59

4	8	7	2	1	5	6	3	9
3	6	1	7	9	4	2	8	5
5	9	2	3	6	8	7	1	4
7	4	5	9	8	2	1	6	3
9	3	6	4	7	1	8	5	2
2	1	8	6	5	3	9	4	7
8	7	3	5	2	6	4	9	1
1	5	9	8	4	7	3	2	6
6	2	4	1	3	9	5	7	8

Solution # 60

7	4	2	6	8	5	1	3	9
1	8	5	3	9	4	2	7	6
6	9	3	2	7	1	8	5	4
4	5	1	8	6	2	7	9	3
9	6	8	7	5	3	4	2	1
2	3	7	1	4	9	5	6	8
5	1	6	4	3	7	9	8	2
8	7	4	9	2	6	3	1	5
3	2	9	5	1	8	6	4	7

Solution # 61

4	3	9	7	1	5	2	8	6
2	8	7	3	9	6	4	5	1
1	5	6	8	4	2	7	3	9
5	4	1	6	8	9	3	2	7
7	6	8	1	2	3	5	9	4
9	2	3	5	7	4	6	1	8
6	9	4	2	5	8	1	7	3
3	7	5	9	6	1	8	4	2
8	1	2	4	3	7	9	6	5

Solution # 62

1	7	5	4	2	6	3	8	9
9	4	2	8	7	3	1	5	6
6	8	3	9	5	1	7	4	2
4	5	9	7	1	2	6	3	8
3	1	8	6	4	9	2	7	5
7	2	6	5	3	8	9	1	4
8	3	4	2	6	7	5	9	1
2	9	7	1	8	5	4	6	3
5	6	1	3	9	4	8	2	7

Solution # 63

2	1	9	5	7	8	3	4	6
8	3	4	9	2	6	7	5	1
7	6	5	4	3	1	9	8	2
9	8	1	7	4	2	5	6	3
6	2	3	1	5	9	8	7	4
4	5	7	6	8	3	1	2	9
5	7	6	3	9	4	2	1	8
1	9	2	8	6	5	4	3	7
3	4	8	2	1	7	6	9	5

Solution # 64

3	8	6	9	7	1	4	2	5
1	7	2	4	3	5	8	6	9
5	4	9	8	2	6	7	1	3
7	2	1	3	6	4	9	5	8
4	6	8	5	1	9	3	7	2
9	3	5	7	8	2	1	4	6
8	5	3	6	4	7	2	9	1
6	1	4	2	9	8	5	3	7
2	9	7	1	5	3	6	8	4

Solution # 65

1	3	8	6	2	4	9	5	7
7	5	6	3	8	9	2	1	4
9	2	4	7	1	5	3	6	8
8	7	3	2	5	1	4	9	6
2	6	5	4	9	7	1	8	3
4	9	1	8	6	3	7	2	5
6	8	7	9	4	2	5	3	1
5	4	9	1	3	6	8	7	2
3	1	2	5	7	8	6	4	9

Solution # 66

5	1	6	8	4	3	9	7	2
9	4	2	7	5	6	1	3	8
8	7	3	1	2	9	4	5	6
1	2	5	4	3	8	6	9	7
6	8	4	5	9	7	3	2	1
7	3	9	2	6	1	8	4	5
4	5	1	3	8	2	7	6	9
3	9	7	6	1	5	2	8	4
2	6	8	9	7	4	5	1	3

Solution # 67

4	1	3	8	6	7	9	5	2
7	2	8	9	3	5	6	1	4
9	6	5	2	1	4	8	7	3
3	4	9	6	7	2	5	8	1
1	8	2	4	5	9	3	6	7
5	7	6	1	8	3	2	4	9
6	3	4	7	2	8	1	9	5
8	5	7	3	9	1	4	2	6
2	9	1	5	4	6	7	3	8

Solution # 68

1	7	5	8	6	4	3	2	9
3	2	8	9	5	1	4	7	6
4	6	9	7	3	2	8	5	1
6	5	4	2	1	3	9	8	7
9	8	2	4	7	6	5	1	3
7	3	1	5	8	9	2	6	4
8	9	3	6	2	7	1	4	5
5	1	7	3	4	8	6	9	2
2	4	6	1	9	5	7	3	8

Solution # 69

9	3	1	5	2	4	8	6	7
6	4	7	9	3	8	1	5	2
8	5	2	7	6	1	3	4	9
4	1	8	2	9	5	7	3	6
2	9	3	6	4	7	5	1	8
7	6	5	8	1	3	9	2	4
1	2	4	3	8	9	6	7	5
5	8	6	1	7	2	4	9	3
3	7	9	4	5	6	2	8	1

Solution # 70

7	3	5	4	8	1	9	2	6
2	9	6	3	7	5	4	1	8
4	1	8	2	6	9	5	7	3
6	7	3	5	4	8	1	9	2
1	5	4	9	3	2	8	6	7
8	2	9	7	1	6	3	5	4
9	4	1	8	2	7	6	3	5
5	8	7	6	9	3	2	4	1
3	6	2	1	5	4	7	8	9

Solution # 71

5	6	1	9	8	2	4	3	7
9	4	2	3	7	1	6	8	5
3	8	7	5	4	6	1	2	9
2	9	5	4	1	8	3	7	6
7	1	4	6	5	3	8	9	2
6	3	8	2	9	7	5	1	4
4	7	3	1	6	9	2	5	8
8	2	6	7	3	5	9	4	1
1	5	9	8	2	4	7	6	3

Solution # 72

6	1	9	2	4	3	8	5	7
4	2	7	5	8	9	6	1	3
3	8	5	6	7	1	2	4	9
5	3	2	1	6	7	4	9	8
8	9	6	3	5	4	7	2	1
7	4	1	9	2	8	5	3	6
9	6	4	7	3	5	1	8	2
2	5	3	8	1	6	9	7	4
1	7	8	4	9	2	3	6	5

Solution # 73

1	3	4	9	6	2	8	7	5
8	2	6	4	5	7	3	1	9
7	5	9	8	3	1	6	2	4
9	1	5	6	7	3	2	4	8
3	8	7	2	1	4	9	5	6
6	4	2	5	8	9	7	3	1
2	9	1	3	4	6	5	8	7
4	6	8	7	2	5	1	9	3
5	7	3	1	9	8	4	6	2

Solution # 74

9	8	5	1	3	2	7	4	6
7	4	2	9	5	6	1	3	8
3	1	6	4	7	8	2	9	5
8	9	3	7	6	5	4	1	2
4	6	1	8	2	9	3	5	7
5	2	7	3	4	1	6	8	9
1	7	9	6	8	4	5	2	3
2	3	8	5	1	7	9	6	4
6	5	4	2	9	3	8	7	1

Solution # 75

5	8	1	7	9	6	4	3	2
2	3	7	1	8	4	6	5	9
6	4	9	5	3	2	8	7	1
4	2	8	9	6	3	5	1	7
7	9	6	8	5	1	3	2	4
3	1	5	2	4	7	9	6	8
9	6	2	3	1	8	7	4	5
1	5	4	6	7	9	2	8	3
8	7	3	4	2	5	1	9	6

Solution # 76

5	9	7	6	1	8	4	3	2
8	2	3	5	4	7	9	6	1
4	1	6	9	3	2	7	8	5
1	6	5	3	7	4	8	2	9
3	7	9	2	8	5	1	4	6
2	4	8	1	9	6	3	5	7
6	3	4	7	5	1	2	9	8
9	5	1	8	2	3	6	7	4
7	8	2	4	6	9	5	1	3

Solution # 77

8	4	1	6	7	9	5	3	2
3	9	2	1	8	5	4	6	7
7	6	5	2	4	3	1	8	9
4	8	3	7	2	1	6	9	5
5	1	9	3	6	4	2	7	8
6	2	7	9	5	8	3	4	1
1	5	4	8	3	7	9	2	6
9	7	6	4	1	2	8	5	3
2	3	8	5	9	6	7	1	4

Solution # 78

5	8	3	6	7	2	4	1	9
7	1	9	3	8	4	2	5	6
6	4	2	5	9	1	8	3	7
3	6	8	9	4	5	1	7	2
2	5	1	8	6	7	9	4	3
4	9	7	2	1	3	6	8	5
8	7	5	4	2	6	3	9	1
9	3	6	1	5	8	7	2	4
1	2	4	7	3	9	5	6	8

Solution # 79

3	9	7	8	6	1	5	2	4
4	8	5	7	3	2	1	6	9
2	6	1	4	5	9	3	7	8
5	7	8	3	1	4	6	9	2
9	4	6	5	2	8	7	1	3
1	2	3	6	9	7	4	8	5
6	5	2	1	8	3	9	4	7
8	3	4	9	7	6	2	5	1
7	1	9	2	4	5	8	3	6

Solution # 80

5	1	3	6	4	8	9	7	2
6	9	7	1	2	3	8	4	5
2	8	4	5	9	7	1	3	6
9	4	8	2	7	5	3	6	1
3	7	2	8	6	1	4	5	9
1	5	6	9	3	4	2	8	7
4	2	1	7	8	6	5	9	3
8	6	5	3	1	9	7	2	4
7	3	9	4	5	2	6	1	8

Solution # 81

2	5	8	9	6	4	1	7	3
9	1	7	2	3	5	8	6	4
6	4	3	7	1	8	2	9	5
3	9	2	8	4	1	6	5	7
1	7	6	5	9	2	3	4	8
4	8	5	3	7	6	9	1	2
8	2	1	4	5	9	7	3	6
7	6	4	1	8	3	5	2	9
5	3	9	6	2	7	4	8	1

Solution # 82

4	8	9	2	1	6	7	3	5
3	2	7	8	4	5	1	6	9
1	6	5	3	7	9	4	8	2
2	5	3	9	6	7	8	1	4
9	7	8	4	3	1	2	5	6
6	4	1	5	8	2	3	9	7
5	1	6	7	2	3	9	4	8
7	3	4	6	9	8	5	2	1
8	9	2	1	5	4	6	7	3

Solution # 83

1	4	8	6	9	5	7	3	2
7	2	9	8	4	3	5	1	6
5	6	3	1	2	7	8	4	9
6	9	7	5	1	4	3	2	8
2	5	4	3	8	9	6	7	1
8	3	1	2	7	6	9	5	4
4	1	5	9	3	8	2	6	7
3	8	2	7	6	1	4	9	5
9	7	6	4	5	2	1	8	3

Solution # 84

2	9	5	7	8	1	6	4	3
8	4	1	6	9	3	7	2	5
3	7	6	5	2	4	8	9	1
6	1	9	4	7	2	5	3	8
4	8	2	3	5	6	9	1	7
7	5	3	8	1	9	2	6	4
9	2	8	1	4	5	3	7	6
1	3	7	2	6	8	4	5	9
5	6	4	9	3	7	1	8	2

Solution # 85

2	8	3	5	4	1	9	6	7
1	4	6	9	7	8	5	2	3
7	5	9	2	3	6	8	4	1
5	2	7	3	1	9	6	8	4
4	9	8	7	6	5	3	1	2
3	6	1	4	8	2	7	5	9
9	3	5	8	2	4	1	7	6
6	7	4	1	5	3	2	9	8
8	1	2	6	9	7	4	3	5

Solution # 86

2	1	9	7	5	4	8	6	3
8	6	5	3	1	9	7	4	2
7	4	3	6	2	8	5	9	1
3	8	2	4	9	1	6	7	5
9	7	1	2	6	5	4	3	8
4	5	6	8	3	7	2	1	9
6	9	8	1	7	2	3	5	4
1	2	7	5	4	3	9	8	6
5	3	4	9	8	6	1	2	7

Solution # 87

6	5	3	7	4	2	8	9	1
4	1	2	6	8	9	3	5	7
7	9	8	3	1	5	2	6	4
3	6	9	2	5	7	1	4	8
5	2	1	8	6	4	7	3	9
8	7	4	9	3	1	6	2	5
9	4	6	1	2	8	5	7	3
2	8	7	5	9	3	4	1	6
1	3	5	4	7	6	9	8	2

Solution # 88

1	8	2	6	4	7	9	3	5
3	9	7	8	1	5	2	6	4
6	4	5	2	3	9	8	7	1
4	6	3	7	2	8	5	1	9
8	2	9	1	5	6	3	4	7
7	5	1	3	9	4	6	8	2
2	1	6	5	7	3	4	9	8
9	7	8	4	6	2	1	5	3
5	3	4	9	8	1	7	2	6

Solution # 89

4	5	1	9	8	2	7	3	6
8	7	3	1	6	5	9	2	4
2	6	9	7	3	4	5	1	8
1	8	4	6	7	3	2	9	5
7	2	6	5	9	1	8	4	3
9	3	5	2	4	8	6	7	1
5	4	7	8	1	9	3	6	2
6	1	8	3	2	7	4	5	9
3	9	2	4	5	6	1	8	7

Solution # 90

5	2	6	7	9	3	8	1	4
8	7	9	4	1	6	5	2	3
1	4	3	2	5	8	6	9	7
3	9	8	1	4	7	2	6	5
6	1	2	5	3	9	4	7	8
4	5	7	8	6	2	1	3	9
9	3	5	6	2	4	7	8	1
2	8	1	9	7	5	3	4	6
7	6	4	3	8	1	9	5	2

Solution # 91

2	8	3	6	5	7	9	1	4
1	4	6	9	8	2	3	5	7
7	9	5	4	3	1	8	6	2
4	6	8	2	7	5	1	9	3
3	7	1	8	6	9	4	2	5
5	2	9	3	1	4	6	7	8
9	5	4	1	2	3	7	8	6
6	3	2	7	9	8	5	4	1
8	1	7	5	4	6	2	3	9

Solution # 92

7	8	2	5	9	6	3	1	4
3	1	9	7	4	2	6	5	8
6	4	5	8	3	1	2	9	7
8	6	7	4	1	5	9	3	2
2	9	4	6	7	3	5	8	1
5	3	1	2	8	9	4	7	6
9	5	8	1	2	4	7	6	3
1	2	3	9	6	7	8	4	5
4	7	6	3	5	8	1	2	9

Solution # 93

3	8	5	7	2	4	1	6	9
7	1	2	9	6	3	8	4	5
9	4	6	5	8	1	3	7	2
5	2	3	6	1	8	4	9	7
1	7	9	3	4	5	6	2	8
4	6	8	2	9	7	5	1	3
2	3	4	8	7	6	9	5	1
6	5	7	1	3	9	2	8	4
8	9	1	4	5	2	7	3	6

Solution # 94

6	3	7	5	1	2	8	9	4
8	2	5	6	4	9	7	1	3
4	9	1	7	3	8	5	2	6
5	1	3	4	9	6	2	7	8
7	4	8	3	2	5	1	6	9
2	6	9	1	8	7	4	3	5
3	7	4	9	5	1	6	8	2
9	8	6	2	7	4	3	5	1
1	5	2	8	6	3	9	4	7

Solution # 95

4	7	9	5	3	1	6	8	2
5	8	6	9	2	4	7	1	3
1	2	3	8	6	7	5	4	9
9	1	2	4	5	6	3	7	8
3	6	4	7	8	2	1	9	5
8	5	7	3	1	9	4	2	6
7	3	8	1	9	5	2	6	4
2	9	1	6	4	3	8	5	7
6	4	5	2	7	8	9	3	1

Solution # 96

1	8	2	9	3	4	6	7	5
4	6	7	2	8	5	9	3	1
5	9	3	7	1	6	8	2	4
6	7	4	5	2	1	3	9	8
8	3	5	6	4	9	2	1	7
2	1	9	3	7	8	5	4	6
3	2	1	8	6	7	4	5	9
7	5	8	4	9	2	1	6	3
9	4	6	1	5	3	7	8	2

Solution # 97

8	1	7	2	5	6	4	3	9
6	2	5	3	9	4	7	8	1
4	9	3	1	7	8	5	2	6
7	5	4	9	2	3	1	6	8
9	6	8	4	1	7	3	5	2
2	3	1	6	8	5	9	7	4
1	8	2	5	3	9	6	4	7
5	7	6	8	4	1	2	9	3
3	4	9	7	6	2	8	1	5

Solution # 98

2	4	1	5	7	9	3	8	6
7	9	8	6	1	3	2	5	4
5	6	3	2	4	8	1	7	9
3	8	7	1	9	4	6	2	5
6	1	9	7	5	2	4	3	8
4	2	5	3	8	6	9	1	7
8	3	6	9	2	7	5	4	1
9	5	4	8	3	1	7	6	2
1	7	2	4	6	5	8	9	3

Solution # 99

1	2	6	8	5	7	9	3	4
9	5	3	2	4	6	8	7	1
8	7	4	9	1	3	5	6	2
4	3	9	5	8	2	6	1	7
5	8	2	6	7	1	4	9	3
6	1	7	4	3	9	2	8	5
2	4	1	3	9	8	7	5	6
3	6	8	7	2	5	1	4	9
7	9	5	1	6	4	3	2	8

Solution # 100

1	6	2	4	7	5	9	3	8
9	4	8	1	6	3	7	2	5
3	7	5	8	9	2	4	1	6
5	9	1	3	4	8	2	6	7
4	2	7	6	5	1	3	8	9
8	3	6	7	2	9	1	5	4
6	5	3	9	1	7	8	4	2
7	1	4	2	8	6	5	9	3
2	8	9	5	3	4	6	7	1

Solution # 101

4	3	5	2	9	6	1	8	7
9	2	6	7	1	8	4	5	3
7	1	8	5	4	3	9	2	6
8	6	2	9	5	1	7	3	4
1	5	4	3	7	2	6	9	8
3	9	7	6	8	4	2	1	5
2	4	9	8	6	5	3	7	1
5	7	1	4	3	9	8	6	2
6	8	3	1	2	7	5	4	9

Solution # 102

8	9	6	7	4	2	3	5	1
1	3	7	6	5	8	9	4	2
2	5	4	9	3	1	8	7	6
6	4	9	2	1	3	7	8	5
5	8	1	4	9	7	2	6	3
3	7	2	5	8	6	1	9	4
4	1	8	3	6	9	5	2	7
7	6	3	8	2	5	4	1	9
9	2	5	1	7	4	6	3	8

Solution # 103

2	4	7	8	5	6	3	1	9
1	5	8	9	3	4	7	2	6
9	6	3	2	7	1	4	5	8
4	7	6	1	9	5	2	8	3
8	1	2	4	6	3	5	9	7
5	3	9	7	8	2	6	4	1
6	9	5	3	2	8	1	7	4
3	8	1	5	4	7	9	6	2
7	2	4	6	1	9	8	3	5

Solution # 104

7	4	6	8	1	3	9	2	5
2	9	3	6	4	5	8	7	1
5	1	8	9	2	7	4	3	6
4	3	9	1	7	2	6	5	8
6	8	7	3	5	9	2	1	4
1	2	5	4	8	6	3	9	7
3	5	4	2	6	1	7	8	9
8	7	2	5	9	4	1	6	3
9	6	1	7	3	8	5	4	2

Solution # 105

6	7	8	3	5	1	2	9	4
4	5	2	8	6	9	7	3	1
3	9	1	2	4	7	6	5	8
8	1	9	5	7	3	4	2	6
5	6	4	9	1	2	8	7	3
2	3	7	6	8	4	9	1	5
9	4	3	1	2	8	5	6	7
7	2	6	4	3	5	1	8	9
1	8	5	7	9	6	3	4	2

Solution # 106

1	5	6	3	2	4	8	7	9
4	9	2	8	7	6	3	5	1
7	8	3	5	9	1	6	2	4
5	3	1	2	8	9	4	6	7
9	2	7	6	4	3	5	1	8
8	6	4	1	5	7	9	3	2
3	4	5	7	1	8	2	9	6
6	7	8	9	3	2	1	4	5
2	1	9	4	6	5	7	8	3

Solution # 107

6	1	2	4	5	7	9	8	3
4	9	8	3	1	2	5	6	7
5	3	7	6	8	9	4	2	1
8	4	3	9	2	5	1	7	6
7	2	9	8	6	1	3	5	4
1	6	5	7	4	3	2	9	8
9	8	1	2	7	4	6	3	5
3	7	4	5	9	6	8	1	2
2	5	6	1	3	8	7	4	9

Solution # 108

2	1	3	7	6	9	8	5	4
8	6	7	4	5	2	1	3	9
4	9	5	3	1	8	2	6	7
3	7	9	6	4	1	5	8	2
1	4	2	5	8	3	7	9	6
5	8	6	9	2	7	3	4	1
6	2	4	1	3	5	9	7	8
7	3	8	2	9	4	6	1	5
9	5	1	8	7	6	4	2	3

Solution # 109

8	9	4	5	3	6	2	1	7
2	1	6	8	7	4	5	9	3
7	3	5	1	2	9	4	6	8
9	7	2	3	4	5	1	8	6
5	8	3	6	1	7	9	2	4
6	4	1	2	9	8	3	7	5
1	2	7	4	6	3	8	5	9
4	5	9	7	8	2	6	3	1
3	6	8	9	5	1	7	4	2

Solution # 110

5	7	6	1	3	8	2	9	4
1	3	9	6	2	4	7	5	8
2	8	4	9	7	5	3	6	1
6	4	7	8	1	9	5	2	3
3	9	5	2	4	6	8	1	7
8	1	2	7	5	3	6	4	9
4	5	1	3	8	2	9	7	6
7	6	3	5	9	1	4	8	2
9	2	8	4	6	7	1	3	5

Solution # 111

8	4	1	7	3	5	9	6	2
5	9	7	6	8	2	3	1	4
6	2	3	1	9	4	8	5	7
9	1	5	3	7	8	2	4	6
2	8	6	4	5	9	1	7	3
3	7	4	2	6	1	5	8	9
7	6	2	5	1	3	4	9	8
1	3	8	9	4	7	6	2	5
4	5	9	8	2	6	7	3	1

Solution # 112

9	4	1	2	8	3	6	5	7
2	5	6	9	1	7	4	8	3
8	7	3	6	5	4	9	2	1
4	3	7	5	9	1	2	6	8
5	6	2	3	4	8	7	1	9
1	9	8	7	2	6	5	3	4
7	2	4	8	3	5	1	9	6
3	1	9	4	6	2	8	7	5
6	8	5	1	7	9	3	4	2

Solution # 113

1	8	2	3	4	6	7	5	9
6	7	4	2	5	9	8	1	3
5	3	9	7	8	1	2	4	6
9	2	6	8	7	4	5	3	1
8	4	5	1	2	3	9	6	7
3	1	7	9	6	5	4	8	2
7	6	1	4	9	8	3	2	5
4	9	3	5	1	2	6	7	8
2	5	8	6	3	7	1	9	4

Solution # 114

6	4	8	9	1	5	2	3	7
9	3	1	2	4	7	5	8	6
7	5	2	6	3	8	1	9	4
4	9	7	8	5	1	3	6	2
2	6	3	7	9	4	8	1	5
8	1	5	3	2	6	7	4	9
3	7	4	1	6	2	9	5	8
1	2	6	5	8	9	4	7	3
5	8	9	4	7	3	6	2	1

Solution # 115

3	1	4	2	7	9	6	8	5
9	7	5	8	4	6	1	3	2
8	2	6	1	3	5	4	9	7
5	6	7	4	9	3	8	2	1
4	8	9	7	2	1	3	5	6
1	3	2	6	5	8	9	7	4
6	5	8	9	1	7	2	4	3
2	9	3	5	6	4	7	1	8
7	4	1	3	8	2	5	6	9

Solution # 116

9	8	5	3	2	6	7	1	4
7	1	2	5	9	4	6	8	3
6	4	3	1	8	7	9	5	2
4	6	1	7	3	8	2	9	5
8	2	9	4	5	1	3	7	6
5	3	7	2	6	9	8	4	1
1	9	8	6	4	2	5	3	7
3	7	6	8	1	5	4	2	9
2	5	4	9	7	3	1	6	8

Solution # 117

1	7	3	9	5	4	8	6	2
8	9	6	1	7	2	3	4	5
4	5	2	8	6	3	7	9	1
2	8	9	6	3	1	5	7	4
3	1	5	7	4	8	6	2	9
6	4	7	5	2	9	1	3	8
5	2	4	3	1	6	9	8	7
9	6	1	4	8	7	2	5	3
7	3	8	2	9	5	4	1	6

Solution # 118

3	9	1	6	2	7	5	8	4
5	7	8	1	3	4	6	2	9
2	4	6	9	5	8	7	1	3
6	2	7	3	8	9	1	4	5
9	5	4	7	1	2	8	3	6
8	1	3	5	4	6	9	7	2
1	8	2	4	9	5	3	6	7
4	6	5	8	7	3	2	9	1
7	3	9	2	6	1	4	5	8

Solution # 119

6	2	1	3	7	4	5	9	8
3	8	4	1	5	9	7	2	6
9	7	5	6	8	2	4	3	1
2	6	3	8	4	7	9	1	5
7	5	8	9	1	6	2	4	3
1	4	9	2	3	5	8	6	7
5	1	2	4	6	8	3	7	9
8	9	6	7	2	3	1	5	4
4	3	7	5	9	1	6	8	2

Solution # 120

8	4	7	2	5	6	1	9	3
1	6	5	3	9	4	7	2	8
3	2	9	8	1	7	6	4	5
5	3	2	1	6	9	4	8	7
6	7	8	4	3	2	9	5	1
9	1	4	5	7	8	2	3	6
2	9	1	6	8	5	3	7	4
4	5	6	7	2	3	8	1	9
7	8	3	9	4	1	5	6	2

Solution # 121

2	9	6	4	1	7	8	3	5
4	8	7	3	9	5	1	2	6
1	3	5	2	6	8	4	9	7
3	7	8	5	4	9	2	6	1
6	1	9	8	2	3	7	5	4
5	4	2	1	7	6	3	8	9
7	5	4	9	3	2	6	1	8
9	6	3	7	8	1	5	4	2
8	2	1	6	5	4	9	7	3

Solution # 122

2	1	9	6	4	8	5	3	7
4	7	3	5	1	9	6	8	2
5	8	6	3	7	2	1	9	4
9	6	4	2	5	7	3	1	8
1	2	7	8	3	6	4	5	9
3	5	8	1	9	4	7	2	6
7	4	1	9	2	3	8	6	5
6	3	2	4	8	5	9	7	1
8	9	5	7	6	1	2	4	3

Solution # 123

2	6	9	3	5	4	8	1	7
5	7	3	8	1	6	9	4	2
8	1	4	9	2	7	6	3	5
6	5	7	4	9	1	2	8	3
3	4	8	7	6	2	5	9	1
9	2	1	5	8	3	4	7	6
4	8	2	1	7	5	3	6	9
7	3	5	6	4	9	1	2	8
1	9	6	2	3	8	7	5	4

Solution # 124

1	2	8	6	3	9	7	5	4
7	5	4	8	2	1	3	9	6
6	9	3	5	7	4	8	2	1
5	3	9	2	8	6	1	4	7
4	7	1	9	5	3	2	6	8
8	6	2	1	4	7	5	3	9
2	4	7	3	9	8	6	1	5
9	1	5	7	6	2	4	8	3
3	8	6	4	1	5	9	7	2

Solution # 125

3	4	2	6	1	7	8	5	9
9	7	1	8	3	5	2	6	4
5	6	8	2	9	4	1	7	3
7	1	3	9	8	2	6	4	5
4	2	5	7	6	3	9	1	8
8	9	6	5	4	1	7	3	2
1	5	9	3	7	8	4	2	6
6	3	7	4	2	9	5	8	1
2	8	4	1	5	6	3	9	7

Solution # 126

8	3	7	2	5	9	4	6	1
2	5	1	8	6	4	7	9	3
4	9	6	1	3	7	8	2	5
1	4	9	7	2	6	5	3	8
5	6	3	4	9	8	1	7	2
7	8	2	5	1	3	9	4	6
3	7	8	6	4	5	2	1	9
9	1	5	3	7	2	6	8	4
6	2	4	9	8	1	3	5	7

Solution # 127

7	8	6	1	3	9	4	2	5
9	5	2	8	4	7	6	3	1
4	3	1	5	6	2	7	9	8
1	9	7	6	2	3	5	8	4
5	4	3	7	9	8	2	1	6
6	2	8	4	5	1	9	7	3
2	7	4	3	8	6	1	5	9
8	6	9	2	1	5	3	4	7
3	1	5	9	7	4	8	6	2

Solution # 128

4	2	7	6	8	9	3	5	1
3	5	8	4	1	7	9	6	2
9	6	1	2	3	5	7	8	4
5	7	4	1	6	3	8	2	9
1	8	3	9	2	4	5	7	6
2	9	6	5	7	8	1	4	3
6	1	5	7	9	2	4	3	8
8	4	2	3	5	1	6	9	7
7	3	9	8	4	6	2	1	5

Solution # 129

5	9	2	3	6	1	8	7	4
1	8	4	7	5	9	3	6	2
3	6	7	2	8	4	1	9	5
8	4	5	1	7	6	2	3	9
7	3	1	4	9	2	6	5	8
9	2	6	8	3	5	7	4	1
4	7	3	5	1	8	9	2	6
2	1	9	6	4	3	5	8	7
6	5	8	9	2	7	4	1	3

Solution # 130

2	9	4	8	5	1	7	6	3
8	5	3	7	2	6	4	9	1
6	7	1	9	3	4	5	2	8
5	2	9	6	8	3	1	4	7
1	6	8	5	4	7	9	3	2
4	3	7	2	1	9	6	8	5
3	1	6	4	7	2	8	5	9
9	8	2	1	6	5	3	7	4
7	4	5	3	9	8	2	1	6

Solution # 131

5	9	2	4	7	8	1	3	6
4	1	7	2	3	6	8	9	5
8	3	6	9	5	1	4	7	2
9	6	8	3	1	2	5	4	7
7	5	3	8	9	4	2	6	1
2	4	1	5	6	7	3	8	9
1	7	5	6	4	3	9	2	8
6	2	4	1	8	9	7	5	3
3	8	9	7	2	5	6	1	4

Solution # 132

9	2	4	1	8	7	3	5	6
7	3	1	6	4	5	9	8	2
6	8	5	9	2	3	4	1	7
5	4	7	2	9	6	8	3	1
3	1	6	4	7	8	2	9	5
2	9	8	5	3	1	7	6	4
1	7	9	3	6	4	5	2	8
8	5	2	7	1	9	6	4	3
4	6	3	8	5	2	1	7	9

Solution # 133

5	4	3	6	2	9	1	8	7
7	8	2	1	3	5	6	9	4
6	1	9	8	7	4	2	3	5
3	2	5	9	4	7	8	6	1
4	6	1	2	8	3	5	7	9
8	9	7	5	6	1	4	2	3
1	3	8	7	5	2	9	4	6
9	7	6	4	1	8	3	5	2
2	5	4	3	9	6	7	1	8

Solution # 134

2	8	5	6	4	7	3	9	1
7	4	6	9	3	1	5	2	8
3	1	9	5	2	8	4	6	7
5	3	4	1	9	2	8	7	6
8	9	2	7	6	3	1	4	5
1	6	7	4	8	5	9	3	2
4	5	1	2	7	9	6	8	3
9	7	8	3	5	6	2	1	4
6	2	3	8	1	4	7	5	9

Solution # 135

1	4	3	5	7	9	2	6	8
2	9	5	6	3	8	7	1	4
8	6	7	1	2	4	9	5	3
6	1	9	2	4	7	8	3	5
3	8	2	9	6	5	1	4	7
7	5	4	8	1	3	6	9	2
5	3	1	7	9	2	4	8	6
9	7	8	4	5	6	3	2	1
4	2	6	3	8	1	5	7	9

Solution # 136

8	9	6	2	1	3	5	7	4
1	2	4	8	7	5	9	3	6
3	7	5	6	9	4	1	8	2
5	3	8	4	2	9	6	1	7
7	1	2	3	6	8	4	9	5
4	6	9	7	5	1	3	2	8
2	5	7	9	3	6	8	4	1
9	8	1	5	4	7	2	6	3
6	4	3	1	8	2	7	5	9

Solution # 137

9	1	3	7	2	8	6	5	4
7	6	5	3	1	4	8	9	2
8	4	2	9	6	5	7	3	1
2	9	1	5	4	7	3	8	6
5	7	6	8	3	2	4	1	9
3	8	4	6	9	1	5	2	7
4	5	9	1	8	6	2	7	3
1	2	8	4	7	3	9	6	5
6	3	7	2	5	9	1	4	8

Solution # 138

9	2	1	8	3	6	4	5	7
4	6	3	5	9	7	1	8	2
7	5	8	2	4	1	3	9	6
5	7	2	9	1	4	6	3	8
3	4	9	6	8	5	7	2	1
1	8	6	7	2	3	9	4	5
2	1	4	3	6	8	5	7	9
8	3	7	1	5	9	2	6	4
6	9	5	4	7	2	8	1	3

Solution # 139

9	4	2	3	1	5	8	6	7
6	7	5	4	9	8	3	1	2
3	8	1	2	6	7	9	4	5
1	9	8	7	4	2	6	5	3
4	5	7	8	3	6	1	2	9
2	3	6	1	5	9	4	7	8
7	1	3	9	2	4	5	8	6
5	2	9	6	8	1	7	3	4
8	6	4	5	7	3	2	9	1

Solution # 140

5	3	2	1	9	4	6	7	8
8	4	9	3	7	6	2	1	5
7	1	6	2	8	5	3	4	9
3	5	4	6	2	7	9	8	1
2	9	1	8	4	3	7	5	6
6	8	7	5	1	9	4	2	3
4	7	3	9	5	8	1	6	2
9	2	8	4	6	1	5	3	7
1	6	5	7	3	2	8	9	4

Solution # 141

9	5	6	1	2	8	7	4	3
4	8	2	7	3	6	5	9	1
3	1	7	9	5	4	2	6	8
7	4	8	2	6	1	9	3	5
5	2	3	4	7	9	8	1	6
1	6	9	3	8	5	4	2	7
2	7	4	5	1	3	6	8	9
6	3	5	8	9	2	1	7	4
8	9	1	6	4	7	3	5	2

Solution # 142

6	1	4	2	8	3	9	5	7
8	9	3	7	5	6	1	2	4
7	5	2	1	4	9	6	3	8
3	7	9	8	6	2	5	4	1
5	4	8	9	7	1	2	6	3
1	2	6	5	3	4	8	7	9
9	8	7	4	2	5	3	1	6
4	6	5	3	1	8	7	9	2
2	3	1	6	9	7	4	8	5

Solution # 143

3	9	7	2	4	6	5	8	1
6	5	2	8	1	7	3	4	9
8	4	1	5	3	9	6	7	2
1	3	6	4	8	5	2	9	7
9	8	5	7	2	3	1	6	4
2	7	4	9	6	1	8	3	5
5	6	3	1	7	4	9	2	8
4	1	8	3	9	2	7	5	6
7	2	9	6	5	8	4	1	3

Solution # 144

4	8	1	5	9	7	3	6	2
7	9	3	2	8	6	4	1	5
6	2	5	1	4	3	9	7	8
9	6	2	4	1	8	7	5	3
3	5	8	7	6	9	1	2	4
1	7	4	3	5	2	6	8	9
5	1	7	8	3	4	2	9	6
2	3	9	6	7	5	8	4	1
8	4	6	9	2	1	5	3	7

Solution # 145

6	1	2	3	4	9	7	8	5
5	8	9	6	2	7	3	1	4
4	3	7	5	8	1	9	2	6
1	4	5	9	7	2	8	6	3
7	6	8	4	3	5	1	9	2
2	9	3	8	1	6	4	5	7
9	5	1	7	6	3	2	4	8
8	7	6	2	9	4	5	3	1
3	2	4	1	5	8	6	7	9

Solution # 146

5	4	3	7	9	6	2	8	1
6	1	8	2	5	3	4	7	9
2	9	7	8	4	1	3	6	5
9	3	2	5	6	8	7	1	4
7	6	1	4	3	2	5	9	8
8	5	4	1	7	9	6	3	2
3	2	6	9	8	5	1	4	7
4	8	5	6	1	7	9	2	3
1	7	9	3	2	4	8	5	6

Solution # 147

7	2	1	8	9	5	4	3	6
9	4	5	2	6	3	1	7	8
3	6	8	4	7	1	2	5	9
2	7	4	6	1	9	3	8	5
8	5	9	3	2	7	6	4	1
1	3	6	5	4	8	7	9	2
6	9	7	1	8	4	5	2	3
5	8	2	7	3	6	9	1	4
4	1	3	9	5	2	8	6	7

Solution # 148

1	7	8	6	9	5	4	3	2
5	4	9	1	3	2	7	8	6
3	6	2	4	7	8	5	1	9
4	9	1	5	2	7	3	6	8
2	8	5	9	6	3	1	4	7
6	3	7	8	4	1	2	9	5
7	2	6	3	1	9	8	5	4
8	1	4	7	5	6	9	2	3
9	5	3	2	8	4	6	7	1

Solution # 149

1	3	9	2	6	5	7	4	8
4	2	5	7	1	8	6	3	9
8	6	7	9	3	4	1	2	5
3	9	4	6	2	1	8	5	7
5	7	2	8	4	9	3	1	6
6	1	8	3	5	7	2	9	4
2	4	3	5	7	6	9	8	1
9	5	6	1	8	2	4	7	3
7	8	1	4	9	3	5	6	2

Solution # 150

3	8	1	7	9	2	5	4	6
6	2	9	5	8	4	7	3	1
4	5	7	1	6	3	2	8	9
2	9	3	4	5	8	6	1	7
1	7	4	2	3	6	8	9	5
5	6	8	9	1	7	3	2	4
8	4	2	6	7	9	1	5	3
9	1	6	3	2	5	4	7	8
7	3	5	8	4	1	9	6	2

Solution # 151

4	7	5	9	1	2	6	8	3
9	1	6	8	3	5	2	4	7
8	2	3	4	7	6	1	5	9
3	5	2	7	4	8	9	1	6
7	6	9	1	5	3	4	2	8
1	4	8	2	6	9	7	3	5
2	9	7	3	8	1	5	6	4
6	8	4	5	2	7	3	9	1
5	3	1	6	9	4	8	7	2

Solution # 152

2	5	8	1	9	3	4	7	6
9	4	7	2	6	8	3	5	1
1	6	3	4	5	7	2	9	8
7	1	9	5	2	6	8	4	3
5	8	2	7	3	4	1	6	9
6	3	4	9	8	1	7	2	5
4	7	5	8	1	9	6	3	2
3	2	1	6	4	5	9	8	7
8	9	6	3	7	2	5	1	4

Solution # 153

3	2	7	5	1	8	9	4	6
4	1	9	6	2	7	5	3	8
5	6	8	3	4	9	7	1	2
6	4	5	9	7	2	1	8	3
7	8	1	4	6	3	2	5	9
9	3	2	1	8	5	4	6	7
1	9	6	7	3	4	8	2	5
8	7	4	2	5	6	3	9	1
2	5	3	8	9	1	6	7	4

Solution # 154

8	1	7	9	2	3	5	4	6
6	9	5	8	1	4	7	3	2
4	2	3	7	6	5	8	1	9
9	3	8	1	7	2	6	5	4
7	6	1	5	4	9	2	8	3
5	4	2	3	8	6	1	9	7
3	7	9	2	5	8	4	6	1
1	8	4	6	3	7	9	2	5
2	5	6	4	9	1	3	7	8

Solution # 155

5	2	3	4	1	7	9	6	8
8	1	9	6	5	3	4	2	7
6	7	4	9	2	8	3	1	5
4	9	5	3	6	1	8	7	2
1	3	7	2	8	4	5	9	6
2	6	8	7	9	5	1	4	3
3	4	1	8	7	2	6	5	9
7	5	6	1	3	9	2	8	4
9	8	2	5	4	6	7	3	1

Solution # 156

8	9	6	1	3	2	7	4	5
7	2	4	5	6	8	9	3	1
3	1	5	4	9	7	8	2	6
5	4	9	2	8	3	1	6	7
1	6	3	7	5	4	2	9	8
2	7	8	9	1	6	4	5	3
6	8	1	3	2	9	5	7	4
9	5	7	6	4	1	3	8	2
4	3	2	8	7	5	6	1	9

Solution # 157

8	1	5	6	7	4	2	3	9
6	2	7	3	9	1	8	4	5
4	9	3	2	5	8	1	7	6
7	4	9	8	6	2	3	5	1
1	8	6	4	3	5	7	9	2
3	5	2	9	1	7	4	6	8
9	3	8	7	2	6	5	1	4
5	6	4	1	8	3	9	2	7
2	7	1	5	4	9	6	8	3

Solution # 158

8	7	6	3	5	2	4	1	9
4	1	2	8	6	9	7	3	5
3	5	9	7	1	4	2	6	8
7	3	5	2	8	1	6	9	4
6	8	4	9	3	5	1	2	7
2	9	1	6	4	7	8	5	3
1	2	3	4	9	8	5	7	6
9	4	7	5	2	6	3	8	1
5	6	8	1	7	3	9	4	2

Solution # 159

5	3	4	6	1	8	9	2	7
1	9	6	5	2	7	3	8	4
8	7	2	9	4	3	6	5	1
2	1	3	4	7	9	8	6	5
4	6	9	8	3	5	7	1	2
7	8	5	2	6	1	4	9	3
9	2	8	3	5	4	1	7	6
6	4	1	7	8	2	5	3	9
3	5	7	1	9	6	2	4	8

Solution # 160

6	3	4	2	9	5	1	7	8
2	1	7	3	4	8	9	5	6
9	8	5	6	7	1	4	2	3
5	2	9	7	6	3	8	4	1
8	4	3	1	5	9	2	6	7
1	7	6	4	8	2	3	9	5
7	6	8	9	3	4	5	1	2
3	9	2	5	1	7	6	8	4
4	5	1	8	2	6	7	3	9

Solution # 161

2	9	1	3	6	5	4	7	8
8	7	6	9	1	4	3	2	5
4	5	3	7	2	8	1	6	9
1	4	9	6	3	2	5	8	7
5	3	7	8	4	9	2	1	6
6	8	2	1	5	7	9	3	4
9	6	5	2	8	3	7	4	1
3	1	4	5	7	6	8	9	2
7	2	8	4	9	1	6	5	3

Solution # 162

7	1	9	2	3	4	6	8	5
3	2	5	6	1	8	9	7	4
6	4	8	7	5	9	1	2	3
9	3	6	4	7	5	8	1	2
5	7	1	8	6	2	4	3	9
2	8	4	1	9	3	5	6	7
1	9	7	5	2	6	3	4	8
4	5	2	3	8	1	7	9	6
8	6	3	9	4	7	2	5	1

Solution # 163

8	4	9	1	7	5	3	6	2
2	1	3	6	8	4	7	9	5
7	5	6	3	9	2	1	8	4
4	2	8	5	3	6	9	1	7
1	3	7	9	2	8	4	5	6
9	6	5	4	1	7	8	2	3
3	8	4	2	6	1	5	7	9
6	9	1	7	5	3	2	4	8
5	7	2	8	4	9	6	3	1

Solution # 164

2	1	9	4	3	5	6	7	8
8	5	3	7	1	6	9	2	4
7	4	6	9	8	2	5	1	3
4	8	2	6	5	3	7	9	1
5	3	7	8	9	1	4	6	2
6	9	1	2	4	7	3	8	5
9	2	5	3	7	8	1	4	6
1	6	4	5	2	9	8	3	7
3	7	8	1	6	4	2	5	9

Solution # 165

1	4	9	8	5	2	7	6	3
8	5	2	6	7	3	9	1	4
3	7	6	1	4	9	8	2	5
5	3	8	7	2	4	1	9	6
6	1	4	9	3	8	2	5	7
2	9	7	5	6	1	4	3	8
4	2	5	3	9	7	6	8	1
9	6	1	4	8	5	3	7	2
7	8	3	2	1	6	5	4	9

Solution # 166

3	7	5	1	6	4	2	8	9
2	9	4	5	3	8	6	7	1
8	1	6	2	7	9	3	4	5
9	2	1	4	8	7	5	3	6
5	4	7	6	2	3	1	9	8
6	3	8	9	5	1	7	2	4
7	6	9	8	1	2	4	5	3
1	8	3	7	4	5	9	6	2
4	5	2	3	9	6	8	1	7

Solution # 167

7	4	1	3	5	2	8	6	9
2	8	3	6	7	9	1	4	5
9	5	6	4	1	8	7	2	3
3	1	9	8	6	7	4	5	2
5	6	7	2	3	4	9	1	8
8	2	4	5	9	1	6	3	7
1	3	2	9	8	6	5	7	4
6	9	5	7	4	3	2	8	1
4	7	8	1	2	5	3	9	6

Solution # 168

3	7	2	9	1	8	5	6	4
8	1	4	6	3	5	7	9	2
5	9	6	4	7	2	1	3	8
2	8	9	5	6	3	4	7	1
4	5	7	8	9	1	6	2	3
6	3	1	2	4	7	8	5	9
9	6	5	1	2	4	3	8	7
7	4	8	3	5	9	2	1	6
1	2	3	7	8	6	9	4	5

Solution # 169

8	2	7	4	6	1	3	9	5
1	5	9	3	2	7	8	4	6
4	3	6	5	8	9	7	1	2
7	1	3	2	4	6	9	5	8
9	4	2	1	5	8	6	7	3
5	6	8	7	9	3	4	2	1
2	8	5	9	3	4	1	6	7
6	7	4	8	1	5	2	3	9
3	9	1	6	7	2	5	8	4

Solution # 170

1	7	2	6	8	4	9	3	5
3	4	9	7	1	5	2	6	8
5	6	8	2	3	9	1	7	4
4	5	1	3	7	6	8	2	9
2	9	6	1	5	8	7	4	3
8	3	7	9	4	2	5	1	6
6	2	3	8	9	7	4	5	1
9	1	4	5	2	3	6	8	7
7	8	5	4	6	1	3	9	2

Solution # 171

3	5	6	2	4	8	9	7	1
9	4	1	3	7	6	5	8	2
8	7	2	1	5	9	3	6	4
2	1	9	5	6	7	4	3	8
6	8	7	4	3	2	1	9	5
5	3	4	8	9	1	7	2	6
1	9	3	6	2	4	8	5	7
7	6	8	9	1	5	2	4	3
4	2	5	7	8	3	6	1	9

Solution # 172

5	1	8	9	6	2	3	4	7
9	4	3	8	5	7	2	6	1
2	6	7	4	3	1	9	5	8
7	2	4	6	1	9	5	8	3
6	3	5	7	4	8	1	2	9
1	8	9	3	2	5	4	7	6
3	5	1	2	8	6	7	9	4
4	7	6	5	9	3	8	1	2
8	9	2	1	7	4	6	3	5

Solution # 173

6	5	4	9	2	3	8	7	1
2	9	1	8	5	7	6	4	3
3	7	8	1	4	6	9	2	5
1	2	9	7	6	5	3	8	4
4	6	7	3	8	9	1	5	2
8	3	5	2	1	4	7	6	9
9	4	2	6	3	8	5	1	7
7	1	6	5	9	2	4	3	8
5	8	3	4	7	1	2	9	6

Solution # 174

5	3	1	6	4	2	9	8	7
8	4	7	9	1	5	3	6	2
6	2	9	8	7	3	4	5	1
4	6	3	2	8	9	1	7	5
7	5	8	3	6	1	2	9	4
9	1	2	4	5	7	6	3	8
3	7	5	1	9	4	8	2	6
2	8	4	7	3	6	5	1	9
1	9	6	5	2	8	7	4	3

Solution # 175

6	9	3	1	5	7	4	2	8
7	4	2	8	6	9	5	1	3
5	1	8	3	4	2	6	7	9
2	3	6	7	1	8	9	5	4
8	5	9	4	2	3	7	6	1
1	7	4	6	9	5	3	8	2
3	6	5	9	8	1	2	4	7
9	2	1	5	7	4	8	3	6
4	8	7	2	3	6	1	9	5

Solution # 176

4	7	3	9	6	1	8	2	5
5	8	6	3	4	2	9	7	1
1	2	9	7	8	5	4	3	6
9	1	7	8	3	6	2	5	4
6	4	8	2	5	7	1	9	3
3	5	2	1	9	4	7	6	8
8	6	1	5	2	9	3	4	7
2	3	4	6	7	8	5	1	9
7	9	5	4	1	3	6	8	2

Solution # 177

8	3	6	1	9	5	7	2	4
5	1	7	8	4	2	6	9	3
4	9	2	3	6	7	1	5	8
3	2	9	7	5	4	8	6	1
6	7	8	2	1	9	3	4	5
1	4	5	6	8	3	2	7	9
7	8	4	5	2	1	9	3	6
9	6	3	4	7	8	5	1	2
2	5	1	9	3	6	4	8	7

Solution # 178

7	3	6	5	1	2	8	4	9
9	8	2	7	6	4	3	5	1
1	5	4	8	9	3	6	7	2
4	2	9	3	5	6	7	1	8
8	7	3	2	4	1	5	9	6
5	6	1	9	7	8	2	3	4
6	4	5	1	8	7	9	2	3
2	9	8	4	3	5	1	6	7
3	1	7	6	2	9	4	8	5

Solution # 179

8	9	3	5	2	7	6	1	4
6	4	2	9	3	1	7	8	5
7	1	5	4	8	6	3	9	2
5	2	9	8	6	4	1	3	7
1	8	4	3	7	2	9	5	6
3	6	7	1	5	9	2	4	8
4	5	6	2	9	3	8	7	1
9	7	8	6	1	5	4	2	3
2	3	1	7	4	8	5	6	9

Solution # 180

5	1	9	7	3	2	8	6	4
8	2	7	4	6	5	9	1	3
3	6	4	8	1	9	2	5	7
2	7	3	6	8	4	5	9	1
4	5	8	3	9	1	7	2	6
6	9	1	5	2	7	3	4	8
1	3	5	2	4	8	6	7	9
7	4	6	9	5	3	1	8	2
9	8	2	1	7	6	4	3	5

Solution # 181

9	8	7	1	4	6	2	5	3
1	3	5	8	7	2	6	9	4
6	2	4	9	3	5	7	8	1
2	4	6	3	5	1	8	7	9
3	5	9	2	8	7	1	4	6
7	1	8	6	9	4	3	2	5
4	9	3	7	6	8	5	1	2
8	6	2	5	1	9	4	3	7
5	7	1	4	2	3	9	6	8

Solution # 182

4	3	5	7	2	1	6	8	9
1	7	6	3	8	9	4	5	2
2	8	9	4	6	5	7	1	3
3	9	7	1	5	2	8	6	4
8	5	4	6	9	7	3	2	1
6	2	1	8	3	4	9	7	5
7	1	3	2	4	8	5	9	6
5	6	2	9	7	3	1	4	8
9	4	8	5	1	6	2	3	7

Solution # 183

5	4	1	2	7	6	8	3	9
9	6	2	4	8	3	5	1	7
7	3	8	1	9	5	6	2	4
3	2	4	6	5	8	7	9	1
8	1	5	9	2	7	3	4	6
6	7	9	3	1	4	2	8	5
1	8	6	5	3	9	4	7	2
2	5	3	7	4	1	9	6	8
4	9	7	8	6	2	1	5	3

Solution # 184

3	6	2	7	8	4	5	1	9
1	5	4	3	2	9	6	8	7
8	7	9	6	5	1	2	4	3
4	9	1	2	7	8	3	5	6
6	2	3	1	9	5	8	7	4
5	8	7	4	6	3	1	9	2
2	3	5	8	4	7	9	6	1
9	4	6	5	1	2	7	3	8
7	1	8	9	3	6	4	2	5

Solution # 185

5	3	1	4	9	8	6	2	7
2	8	9	7	6	1	3	4	5
6	7	4	2	3	5	9	1	8
8	4	6	1	7	9	5	3	2
7	9	5	3	8	2	1	6	4
1	2	3	5	4	6	7	8	9
4	6	7	8	5	3	2	9	1
9	5	2	6	1	4	8	7	3
3	1	8	9	2	7	4	5	6

Solution # 186

6	4	7	2	1	5	3	8	9
5	3	1	4	8	9	2	7	6
8	2	9	6	3	7	5	4	1
2	1	8	3	7	6	9	5	4
9	6	3	8	5	4	7	1	2
4	7	5	1	9	2	6	3	8
1	9	6	5	4	3	8	2	7
7	5	4	9	2	8	1	6	3
3	8	2	7	6	1	4	9	5

Solution # 187

1	4	3	2	6	7	9	8	5
7	8	6	9	5	1	2	3	4
5	9	2	4	8	3	6	7	1
8	7	9	1	3	6	4	5	2
2	1	5	7	4	8	3	6	9
3	6	4	5	2	9	8	1	7
4	3	7	8	9	5	1	2	6
6	2	1	3	7	4	5	9	8
9	5	8	6	1	2	7	4	3

Solution # 188

7	1	9	8	5	3	4	6	2
6	5	8	1	4	2	3	7	9
4	3	2	9	7	6	8	1	5
2	6	3	4	1	8	5	9	7
1	7	4	5	2	9	6	3	8
9	8	5	3	6	7	2	4	1
5	4	6	7	8	1	9	2	3
8	9	7	2	3	4	1	5	6
3	2	1	6	9	5	7	8	4

Solution # 189

6	5	7	1	2	4	8	3	9
2	8	9	3	7	6	1	4	5
3	1	4	9	8	5	6	7	2
9	4	3	5	6	2	7	1	8
1	7	2	8	4	3	9	5	6
8	6	5	7	1	9	4	2	3
4	9	6	2	3	7	5	8	1
5	3	1	4	9	8	2	6	7
7	2	8	6	5	1	3	9	4

Solution # 190

8	7	1	5	3	4	6	2	9
4	6	9	8	2	7	5	1	3
2	5	3	1	6	9	4	7	8
3	8	5	6	7	2	9	4	1
6	1	7	9	4	8	2	3	5
9	4	2	3	1	5	8	6	7
5	2	6	7	8	3	1	9	4
1	3	8	4	9	6	7	5	2
7	9	4	2	5	1	3	8	6

Solution # 191

4	2	8	5	1	7	6	9	3
3	5	7	8	9	6	2	4	1
1	9	6	2	3	4	5	7	8
6	8	5	7	4	9	3	1	2
9	4	2	3	5	1	8	6	7
7	3	1	6	2	8	9	5	4
2	7	9	4	6	3	1	8	5
5	6	4	1	8	2	7	3	9
8	1	3	9	7	5	4	2	6

Solution # 192

5	9	1	4	2	7	3	6	8
2	6	4	3	9	8	1	7	5
7	3	8	5	6	1	9	4	2
1	2	9	6	8	3	4	5	7
6	4	7	1	5	2	8	3	9
3	8	5	9	7	4	2	1	6
8	7	3	2	1	6	5	9	4
4	5	2	7	3	9	6	8	1
9	1	6	8	4	5	7	2	3

Solution # 193

2	7	5	3	8	6	1	9	4
4	8	9	1	7	5	3	6	2
1	6	3	2	9	4	5	8	7
6	1	4	5	3	7	9	2	8
9	3	2	8	4	1	7	5	6
8	5	7	6	2	9	4	1	3
5	2	6	7	1	3	8	4	9
3	9	8	4	5	2	6	7	1
7	4	1	9	6	8	2	3	5

Solution # 194

7	3	9	8	5	4	1	6	2
6	8	4	1	7	2	3	9	5
1	5	2	3	6	9	4	8	7
8	2	5	7	3	6	9	4	1
9	7	3	2	4	1	8	5	6
4	6	1	5	9	8	2	7	3
3	4	8	6	1	7	5	2	9
2	1	7	9	8	5	6	3	4
5	9	6	4	2	3	7	1	8

Solution # 195

7	4	2	9	1	8	5	3	6
5	6	9	2	3	7	8	4	1
3	8	1	6	4	5	9	7	2
4	9	7	5	2	6	3	1	8
1	2	8	4	7	3	6	9	5
6	3	5	1	8	9	4	2	7
9	5	3	7	6	1	2	8	4
8	7	4	3	5	2	1	6	9
2	1	6	8	9	4	7	5	3

Solution # 196

6	7	2	4	8	3	5	9	1
4	3	9	1	5	2	6	8	7
5	1	8	9	6	7	2	4	3
1	6	5	2	7	4	9	3	8
8	2	4	5	3	9	7	1	6
7	9	3	6	1	8	4	5	2
3	4	7	8	2	5	1	6	9
2	5	1	3	9	6	8	7	4
9	8	6	7	4	1	3	2	5

Solution # 197

9	3	6	8	2	5	7	1	4
1	2	5	4	7	9	6	3	8
8	7	4	3	1	6	9	5	2
7	6	8	1	4	3	5	2	9
2	1	9	5	6	8	3	4	7
5	4	3	7	9	2	1	8	6
6	8	7	2	5	1	4	9	3
3	9	1	6	8	4	2	7	5
4	5	2	9	3	7	8	6	1

Solution # 198

9	7	2	1	4	5	8	6	3
5	3	1	2	6	8	4	9	7
6	8	4	9	7	3	2	5	1
8	5	3	4	2	9	1	7	6
4	2	7	3	1	6	9	8	5
1	9	6	5	8	7	3	4	2
2	4	9	6	5	1	7	3	8
7	1	5	8	3	4	6	2	9
3	6	8	7	9	2	5	1	4

Solution # 199

1	3	5	6	9	2	8	4	7
4	2	6	7	3	8	9	5	1
7	9	8	4	1	5	2	3	6
9	5	3	2	6	7	1	8	4
6	8	4	1	5	9	3	7	2
2	1	7	8	4	3	5	6	9
8	7	1	3	2	4	6	9	5
3	6	9	5	7	1	4	2	8
5	4	2	9	8	6	7	1	3

Solution # 200

9	3	1	8	4	6	5	2	7
8	4	5	3	7	2	9	1	6
2	7	6	9	1	5	4	3	8
6	9	3	1	5	8	7	4	2
5	8	7	4	2	9	1	6	3
1	2	4	7	6	3	8	5	9
7	6	2	5	8	1	3	9	4
4	5	9	6	3	7	2	8	1
3	1	8	2	9	4	6	7	5

Solution # 201

2	1	5	9	7	3	8	4	6
7	6	3	4	8	5	9	2	1
9	8	4	1	2	6	7	3	5
8	3	1	6	9	4	5	7	2
5	9	7	2	3	1	6	8	4
6	4	2	7	5	8	1	9	3
1	7	9	5	4	2	3	6	8
3	2	6	8	1	7	4	5	9
4	5	8	3	6	9	2	1	7

Solution # 202

4	6	5	8	3	9	1	2	7
9	3	2	1	7	6	4	5	8
7	1	8	2	4	5	6	3	9
5	4	3	6	8	7	2	9	1
1	8	6	5	9	2	7	4	3
2	9	7	4	1	3	8	6	5
8	5	1	3	6	4	9	7	2
3	7	4	9	2	8	5	1	6
6	2	9	7	5	1	3	8	4

Solution # 203

8	1	7	3	2	6	4	9	5
4	3	2	9	1	5	6	7	8
5	9	6	4	7	8	2	3	1
9	8	4	6	3	2	5	1	7
1	6	3	5	9	7	8	2	4
2	7	5	8	4	1	9	6	3
6	4	9	1	5	3	7	8	2
3	2	8	7	6	4	1	5	9
7	5	1	2	8	9	3	4	6

Solution # 204

3	7	9	5	6	8	1	4	2
5	1	8	2	4	3	9	7	6
2	6	4	7	9	1	5	8	3
8	4	6	9	5	2	3	1	7
1	2	3	8	7	6	4	9	5
9	5	7	1	3	4	2	6	8
4	8	5	3	1	7	6	2	9
7	9	1	6	2	5	8	3	4
6	3	2	4	8	9	7	5	1

Solution # 205

2	1	5	6	7	4	3	8	9
4	7	3	9	8	2	1	5	6
6	9	8	1	5	3	4	7	2
1	8	6	5	3	9	2	4	7
5	2	9	7	4	1	8	6	3
7	3	4	8	2	6	9	1	5
8	5	2	3	1	7	6	9	4
3	6	1	4	9	5	7	2	8
9	4	7	2	6	8	5	3	1

Solution # 206

6	7	2	3	8	4	1	9	5
4	5	9	1	6	2	8	3	7
3	1	8	7	9	5	2	4	6
9	6	5	4	1	8	3	7	2
1	4	3	5	2	7	6	8	9
2	8	7	9	3	6	5	1	4
5	3	1	6	4	9	7	2	8
7	2	4	8	5	3	9	6	1
8	9	6	2	7	1	4	5	3

Solution # 207

1	7	9	3	8	2	5	6	4
4	5	6	1	9	7	2	3	8
3	2	8	5	6	4	1	7	9
8	3	4	6	2	9	7	1	5
5	9	1	7	4	3	8	2	6
2	6	7	8	5	1	4	9	3
7	1	5	9	3	8	6	4	2
9	8	2	4	7	6	3	5	1
6	4	3	2	1	5	9	8	7

Solution # 208

5	2	4	7	6	9	3	8	1
3	1	6	8	5	4	2	9	7
8	7	9	3	2	1	4	5	6
9	6	5	2	8	3	7	1	4
4	3	7	1	9	6	8	2	5
1	8	2	4	7	5	6	3	9
6	4	3	9	1	8	5	7	2
7	5	1	6	3	2	9	4	8
2	9	8	5	4	7	1	6	3

Solution # 209

1	7	5	2	6	4	3	8	9
3	9	4	8	1	7	6	2	5
8	6	2	5	9	3	1	7	4
6	5	3	4	7	2	9	1	8
9	4	1	6	3	8	7	5	2
7	2	8	1	5	9	4	3	6
4	1	9	3	2	5	8	6	7
2	3	7	9	8	6	5	4	1
5	8	6	7	4	1	2	9	3

Solution # 210

9	4	1	7	3	2	6	8	5
2	7	8	6	1	5	3	4	9
5	6	3	9	4	8	1	7	2
7	3	6	1	8	9	2	5	4
4	8	2	5	7	3	9	1	6
1	9	5	4	2	6	8	3	7
3	5	7	2	9	1	4	6	8
6	1	9	8	5	4	7	2	3
8	2	4	3	6	7	5	9	1

Solution # 211

6	9	5	8	2	3	1	4	7
4	3	7	9	1	5	6	2	8
2	1	8	6	7	4	5	9	3
8	2	9	7	5	1	3	6	4
5	7	6	4	3	9	2	8	1
1	4	3	2	6	8	7	5	9
7	8	1	5	9	2	4	3	6
9	6	2	3	4	7	8	1	5
3	5	4	1	8	6	9	7	2

Solution # 212

9	3	8	6	4	5	2	7	1
7	4	6	3	2	1	8	9	5
2	5	1	9	8	7	4	6	3
6	8	9	1	5	4	7	3	2
3	7	5	8	9	2	1	4	6
4	1	2	7	3	6	9	5	8
8	9	7	2	6	3	5	1	4
1	6	4	5	7	8	3	2	9
5	2	3	4	1	9	6	8	7

Solution # 213

4	6	9	2	8	7	5	3	1
5	2	3	1	9	6	4	7	8
8	7	1	3	5	4	6	9	2
7	4	5	9	6	8	2	1	3
9	1	6	5	2	3	8	4	7
3	8	2	7	4	1	9	5	6
2	3	4	6	1	9	7	8	5
1	5	8	4	7	2	3	6	9
6	9	7	8	3	5	1	2	4

Solution # 214

9	2	1	4	6	7	5	3	8
8	7	4	3	5	2	6	1	9
5	3	6	9	1	8	4	7	2
6	1	8	5	7	3	2	9	4
4	5	3	2	9	6	7	8	1
7	9	2	1	8	4	3	5	6
1	6	9	7	2	5	8	4	3
2	4	5	8	3	9	1	6	7
3	8	7	6	4	1	9	2	5

Solution # 215

3	4	9	5	1	8	2	6	7
6	5	8	9	2	7	4	3	1
1	7	2	3	4	6	5	9	8
7	3	4	2	5	9	8	1	6
2	9	5	8	6	1	3	7	4
8	6	1	7	3	4	9	5	2
5	2	7	1	8	3	6	4	9
9	8	6	4	7	5	1	2	3
4	1	3	6	9	2	7	8	5

Solution # 216

9	3	8	6	4	5	1	2	7
7	2	6	3	8	1	9	5	4
4	1	5	7	2	9	3	6	8
5	7	1	4	9	6	2	8	3
2	8	4	1	7	3	6	9	5
3	6	9	2	5	8	4	7	1
8	9	2	5	3	4	7	1	6
1	4	7	8	6	2	5	3	9
6	5	3	9	1	7	8	4	2

Solution # 217

7	5	2	9	1	4	6	3	8
3	4	9	8	6	2	7	1	5
6	8	1	3	5	7	9	4	2
8	2	3	1	9	6	4	5	7
1	7	5	4	3	8	2	6	9
4	9	6	7	2	5	3	8	1
5	6	4	2	7	1	8	9	3
2	3	8	5	4	9	1	7	6
9	1	7	6	8	3	5	2	4

Solution # 218

7	8	3	4	9	1	2	6	5
6	1	4	7	5	2	3	8	9
2	5	9	3	6	8	4	1	7
5	6	2	8	7	9	1	3	4
3	7	8	6	1	4	9	5	2
4	9	1	2	3	5	8	7	6
1	3	7	9	2	6	5	4	8
9	4	6	5	8	3	7	2	1
8	2	5	1	4	7	6	9	3

Solution # 219

1	3	8	2	6	4	7	9	5
7	5	6	3	9	1	8	4	2
9	2	4	5	8	7	1	3	6
2	8	3	4	1	9	6	5	7
6	7	5	8	3	2	9	1	4
4	1	9	7	5	6	2	8	3
3	9	2	1	7	5	4	6	8
5	4	1	6	2	8	3	7	9
8	6	7	9	4	3	5	2	1

Solution # 220

4	5	1	3	9	8	2	6	7
3	2	9	5	6	7	1	8	4
7	6	8	1	4	2	5	3	9
5	3	4	9	7	1	8	2	6
6	1	2	4	8	5	9	7	3
9	8	7	6	2	3	4	5	1
8	4	6	7	5	9	3	1	2
2	7	3	8	1	4	6	9	5
1	9	5	2	3	6	7	4	8

Solution # 221

7	5	2	4	1	8	3	6	9
9	4	3	6	7	5	1	2	8
1	8	6	2	3	9	4	7	5
8	2	7	3	5	1	9	4	6
4	1	5	9	8	6	2	3	7
6	3	9	7	4	2	5	8	1
5	7	1	8	2	4	6	9	3
3	6	4	5	9	7	8	1	2
2	9	8	1	6	3	7	5	4

Solution # 222

1	2	8	9	7	4	3	5	6
5	3	7	8	6	2	4	9	1
6	4	9	3	1	5	8	7	2
4	8	6	7	2	9	1	3	5
3	9	5	6	8	1	7	2	4
2	7	1	5	4	3	6	8	9
8	5	3	4	9	6	2	1	7
7	6	2	1	5	8	9	4	3
9	1	4	2	3	7	5	6	8

Solution # 223

6	3	8	7	1	9	2	4	5
5	7	2	4	6	3	1	8	9
1	9	4	2	8	5	7	6	3
3	5	6	9	2	4	8	7	1
7	2	1	8	5	6	3	9	4
8	4	9	3	7	1	6	5	2
2	6	3	5	4	7	9	1	8
4	8	7	1	9	2	5	3	6
9	1	5	6	3	8	4	2	7

Solution # 224

6	4	2	7	5	3	8	9	1
9	1	7	4	8	6	2	5	3
5	3	8	2	1	9	7	6	4
3	5	1	8	9	2	4	7	6
8	9	6	3	7	4	1	2	5
7	2	4	1	6	5	9	3	8
4	8	5	9	3	7	6	1	2
2	7	3	6	4	1	5	8	9
1	6	9	5	2	8	3	4	7

Solution # 225

8	1	5	7	2	3	6	4	9
3	4	9	1	8	6	2	7	5
7	2	6	5	4	9	3	8	1
4	5	1	9	6	8	7	2	3
6	3	8	2	7	1	5	9	4
2	9	7	3	5	4	1	6	8
5	6	4	8	1	7	9	3	2
1	7	3	4	9	2	8	5	6
9	8	2	6	3	5	4	1	7

Solution # 226

7	1	4	2	6	8	5	3	9
2	6	8	5	3	9	4	1	7
9	3	5	7	4	1	8	2	6
6	7	2	9	1	4	3	5	8
1	4	9	3	8	5	6	7	2
5	8	3	6	2	7	9	4	1
3	2	1	4	9	6	7	8	5
4	5	6	8	7	2	1	9	3
8	9	7	1	5	3	2	6	4

Solution # 227

6	9	4	7	3	8	5	1	2
8	2	5	9	6	1	7	3	4
1	7	3	4	5	2	6	9	8
3	4	7	8	9	5	1	2	6
2	6	9	1	7	3	8	4	5
5	8	1	6	2	4	3	7	9
7	5	8	2	1	9	4	6	3
4	1	2	3	8	6	9	5	7
9	3	6	5	4	7	2	8	1

Solution # 228

8	4	3	7	9	6	2	1	5
5	1	9	3	2	8	7	4	6
7	6	2	4	5	1	8	3	9
3	5	4	6	8	7	1	9	2
2	8	6	5	1	9	4	7	3
1	9	7	2	3	4	6	5	8
6	3	8	1	7	5	9	2	4
4	7	5	9	6	2	3	8	1
9	2	1	8	4	3	5	6	7

Solution # 229

7	3	9	5	1	6	4	2	8
2	6	1	8	4	9	5	3	7
8	5	4	7	2	3	9	6	1
6	7	5	9	8	4	3	1	2
9	2	8	3	7	1	6	5	4
1	4	3	6	5	2	7	8	9
4	8	6	1	9	5	2	7	3
3	1	2	4	6	7	8	9	5
5	9	7	2	3	8	1	4	6

Solution # 230

6	2	5	9	1	3	4	8	7
1	8	4	6	7	5	2	3	9
3	7	9	2	8	4	1	5	6
9	4	1	7	5	2	8	6	3
2	6	8	4	3	9	5	7	1
7	5	3	8	6	1	9	4	2
4	1	7	5	2	6	3	9	8
8	9	2	3	4	7	6	1	5
5	3	6	1	9	8	7	2	4

Solution # 231

6	1	4	5	7	2	8	9	3
7	3	5	1	9	8	2	4	6
9	2	8	3	6	4	5	7	1
8	9	7	6	5	3	1	2	4
2	5	1	7	4	9	6	3	8
4	6	3	2	8	1	9	5	7
3	8	9	4	2	6	7	1	5
1	7	2	8	3	5	4	6	9
5	4	6	9	1	7	3	8	2

Solution # 232

9	7	4	1	3	5	8	6	2
6	8	3	7	2	4	1	5	9
1	2	5	6	9	8	7	4	3
8	3	1	9	6	2	5	7	4
7	4	2	5	8	3	6	9	1
5	9	6	4	7	1	2	3	8
3	5	8	2	4	7	9	1	6
2	1	9	3	5	6	4	8	7
4	6	7	8	1	9	3	2	5

Solution # 233

2	3	7	4	1	6	5	9	8
5	6	9	3	8	2	1	7	4
1	8	4	5	9	7	2	3	6
8	1	6	7	2	4	3	5	9
4	9	3	1	6	5	8	2	7
7	2	5	8	3	9	6	4	1
3	7	1	2	4	8	9	6	5
9	4	2	6	5	1	7	8	3
6	5	8	9	7	3	4	1	2

Solution # 234

1	5	8	9	6	7	3	4	2
9	3	2	5	8	4	7	6	1
4	6	7	3	1	2	8	5	9
5	9	6	4	3	8	2	1	7
8	4	3	2	7	1	5	9	6
2	7	1	6	5	9	4	8	3
6	1	5	7	4	3	9	2	8
3	2	4	8	9	6	1	7	5
7	8	9	1	2	5	6	3	4

Solution # 235

6	7	8	3	9	5	4	2	1
9	4	3	1	7	2	6	5	8
1	2	5	6	8	4	9	7	3
3	1	2	9	6	8	7	4	5
4	5	6	7	2	1	8	3	9
7	8	9	5	4	3	1	6	2
8	3	4	2	1	6	5	9	7
2	6	7	8	5	9	3	1	4
5	9	1	4	3	7	2	8	6

Solution # 236

2	6	7	3	4	9	8	5	1
5	8	9	2	7	1	4	3	6
3	4	1	6	5	8	7	9	2
6	3	5	9	8	7	1	2	4
8	7	4	1	2	5	9	6	3
9	1	2	4	3	6	5	8	7
4	9	6	8	1	3	2	7	5
7	2	3	5	9	4	6	1	8
1	5	8	7	6	2	3	4	9

Solution # 237

7	3	8	1	2	5	4	9	6
2	9	1	3	4	6	5	8	7
6	4	5	9	8	7	1	2	3
5	1	4	6	7	2	8	3	9
3	2	7	4	9	8	6	5	1
8	6	9	5	3	1	7	4	2
4	7	6	2	5	9	3	1	8
1	5	2	8	6	3	9	7	4
9	8	3	7	1	4	2	6	5

Solution # 238

9	5	6	2	8	4	3	1	7
2	7	3	1	6	9	5	8	4
4	8	1	7	3	5	9	6	2
7	1	9	3	2	6	4	5	8
3	2	8	4	5	1	7	9	6
6	4	5	9	7	8	1	2	3
5	3	7	8	9	2	6	4	1
8	6	4	5	1	3	2	7	9
1	9	2	6	4	7	8	3	5

Solution # 239

5	2	3	4	8	7	1	9	6
4	8	9	1	2	6	3	7	5
6	7	1	3	5	9	2	8	4
2	5	8	7	1	4	9	6	3
7	3	6	8	9	5	4	1	2
9	1	4	2	6	3	7	5	8
8	9	2	6	3	1	5	4	7
1	6	7	5	4	2	8	3	9
3	4	5	9	7	8	6	2	1

Solution # 240

1	3	2	9	5	7	8	6	4
7	8	5	3	6	4	9	2	1
6	9	4	1	8	2	7	3	5
9	5	6	8	2	1	4	7	3
4	1	8	6	7	3	2	5	9
3	2	7	5	4	9	1	8	6
5	4	1	7	3	8	6	9	2
2	7	3	4	9	6	5	1	8
8	6	9	2	1	5	3	4	7

Solution # 241

5	7	4	6	1	2	9	8	3
1	3	9	8	4	5	7	6	2
2	8	6	7	3	9	1	4	5
8	9	7	4	6	3	5	2	1
4	1	2	5	8	7	3	9	6
3	6	5	9	2	1	4	7	8
7	4	8	3	5	6	2	1	9
6	2	3	1	9	4	8	5	7
9	5	1	2	7	8	6	3	4

Solution # 242

9	3	8	4	5	1	2	7	6
7	2	4	9	6	3	5	8	1
1	6	5	7	2	8	3	9	4
4	1	2	5	8	9	6	3	7
3	8	7	6	1	4	9	2	5
6	5	9	3	7	2	4	1	8
2	7	3	1	4	6	8	5	9
8	4	1	2	9	5	7	6	3
5	9	6	8	3	7	1	4	2

Solution # 243

5	1	3	8	6	7	2	9	4
7	6	8	4	2	9	3	5	1
9	2	4	5	1	3	6	8	7
4	3	2	1	9	6	5	7	8
6	8	5	7	3	4	1	2	9
1	7	9	2	5	8	4	3	6
8	5	7	6	4	2	9	1	3
3	4	1	9	8	5	7	6	2
2	9	6	3	7	1	8	4	5

Solution # 244

7	8	1	4	9	5	2	6	3
9	2	6	3	1	8	7	5	4
3	5	4	2	7	6	8	1	9
2	4	5	1	8	7	9	3	6
1	9	3	6	5	2	4	7	8
6	7	8	9	3	4	1	2	5
5	3	2	8	4	1	6	9	7
4	1	7	5	6	9	3	8	2
8	6	9	7	2	3	5	4	1

Solution # 245

4	7	1	2	6	5	8	3	9
9	6	8	7	3	4	1	2	5
3	2	5	9	8	1	7	6	4
8	3	2	5	4	9	6	1	7
5	1	9	6	7	3	4	8	2
7	4	6	1	2	8	9	5	3
1	9	3	4	5	6	2	7	8
6	8	7	3	9	2	5	4	1
2	5	4	8	1	7	3	9	6

Solution # 246

5	9	8	6	2	3	7	4	1
7	2	4	1	5	8	9	6	3
3	6	1	7	4	9	5	8	2
6	4	7	5	3	2	8	1	9
1	5	3	8	9	6	2	7	4
9	8	2	4	1	7	6	3	5
8	3	9	2	7	1	4	5	6
2	7	5	3	6	4	1	9	8
4	1	6	9	8	5	3	2	7

Solution # 247

8	2	4	5	9	6	3	7	1
9	7	6	1	2	3	5	8	4
3	1	5	4	8	7	9	2	6
5	8	9	7	3	1	6	4	2
4	3	1	9	6	2	7	5	8
7	6	2	8	5	4	1	9	3
6	4	3	2	7	9	8	1	5
2	9	8	6	1	5	4	3	7
1	5	7	3	4	8	2	6	9

Solution # 248

9	6	4	2	5	8	7	3	1
1	5	3	6	7	9	8	4	2
7	8	2	4	3	1	6	9	5
3	7	9	5	4	2	1	8	6
6	2	1	8	9	7	3	5	4
5	4	8	1	6	3	9	2	7
8	3	6	7	2	5	4	1	9
4	9	5	3	1	6	2	7	8
2	1	7	9	8	4	5	6	3

Solution # 249

5	8	4	7	2	6	1	3	9
7	9	6	3	4	1	2	8	5
3	2	1	8	5	9	7	6	4
1	5	7	9	3	4	6	2	8
4	3	2	6	7	8	9	5	1
8	6	9	2	1	5	4	7	3
6	7	5	1	9	3	8	4	2
2	1	3	4	8	7	5	9	6
9	4	8	5	6	2	3	1	7

Solution # 250

9	6	3	2	4	7	5	1	8
7	8	2	3	5	1	4	9	6
5	1	4	9	6	8	2	3	7
6	2	9	5	8	3	7	4	1
1	3	7	6	2	4	9	8	5
4	5	8	7	1	9	3	6	2
2	9	5	1	3	6	8	7	4
3	4	6	8	7	2	1	5	9
8	7	1	4	9	5	6	2	3

Solution # 251

5	9	8	7	1	2	6	4	3
4	1	3	8	6	9	5	2	7
2	7	6	3	5	4	1	8	9
3	5	9	6	8	1	2	7	4
1	2	7	4	9	5	3	6	8
8	6	4	2	7	3	9	5	1
7	8	2	9	3	6	4	1	5
9	4	5	1	2	7	8	3	6
6	3	1	5	4	8	7	9	2

Solution # 252

6	8	1	9	3	7	4	2	5
2	3	5	1	4	6	7	8	9
4	9	7	2	8	5	6	3	1
5	2	8	6	9	4	3	1	7
9	1	4	8	7	3	2	5	6
7	6	3	5	1	2	8	9	4
8	4	9	3	6	1	5	7	2
3	7	2	4	5	9	1	6	8
1	5	6	7	2	8	9	4	3

Solution # 253

9	5	8	1	3	2	4	7	6
7	2	4	9	5	6	8	1	3
6	1	3	4	7	8	5	9	2
2	8	6	7	1	4	3	5	9
3	7	1	5	8	9	6	2	4
4	9	5	2	6	3	7	8	1
1	3	9	8	4	5	2	6	7
8	6	7	3	2	1	9	4	5
5	4	2	6	9	7	1	3	8

Solution # 254

6	5	9	1	8	3	7	2	4
2	3	4	7	6	9	1	5	8
7	1	8	4	5	2	3	9	6
8	9	1	6	4	7	5	3	2
4	6	7	2	3	5	9	8	1
5	2	3	9	1	8	4	6	7
1	8	5	3	7	6	2	4	9
3	7	2	8	9	4	6	1	5
9	4	6	5	2	1	8	7	3

Solution # 255

6	3	2	9	8	5	4	1	7
1	5	7	4	2	6	3	9	8
4	8	9	3	1	7	6	5	2
5	7	1	8	4	9	2	6	3
3	4	6	2	5	1	7	8	9
2	9	8	7	6	3	1	4	5
8	2	5	1	3	4	9	7	6
9	6	4	5	7	2	8	3	1
7	1	3	6	9	8	5	2	4

Solution # 256

2	6	7	5	9	3	1	8	4
5	8	1	4	7	2	3	9	6
3	4	9	6	8	1	2	5	7
6	7	3	8	2	9	4	1	5
9	2	5	1	3	4	7	6	8
8	1	4	7	5	6	9	2	3
7	9	2	3	6	5	8	4	1
4	5	8	2	1	7	6	3	9
1	3	6	9	4	8	5	7	2

Solution # 257

9	7	6	4	5	2	3	1	8
5	4	1	9	8	3	6	7	2
3	8	2	6	1	7	5	9	4
4	6	5	8	7	1	9	2	3
7	9	8	3	2	6	4	5	1
1	2	3	5	9	4	7	8	6
6	5	9	2	3	8	1	4	7
8	1	4	7	6	5	2	3	9
2	3	7	1	4	9	8	6	5

Solution # 258

1	3	7	5	2	9	8	4	6
6	8	5	3	4	1	7	9	2
4	9	2	7	8	6	3	5	1
7	5	9	8	3	2	1	6	4
8	1	4	6	7	5	9	2	3
2	6	3	9	1	4	5	8	7
3	2	1	4	9	8	6	7	5
9	4	6	1	5	7	2	3	8
5	7	8	2	6	3	4	1	9

Solution # 259

9	6	1	7	3	5	8	2	4
2	3	4	9	6	8	5	7	1
8	5	7	1	2	4	3	6	9
1	7	2	8	9	3	6	4	5
4	8	3	6	5	1	7	9	2
5	9	6	4	7	2	1	3	8
3	1	8	2	4	6	9	5	7
7	2	5	3	1	9	4	8	6
6	4	9	5	8	7	2	1	3

Solution # 260

3	1	9	5	2	6	7	4	8
4	6	7	3	8	9	5	2	1
8	5	2	1	7	4	6	3	9
6	3	1	8	9	5	4	7	2
7	2	8	4	6	3	1	9	5
9	4	5	2	1	7	8	6	3
5	9	3	6	4	1	2	8	7
2	7	4	9	5	8	3	1	6
1	8	6	7	3	2	9	5	4

Solution # 261

1	3	2	5	4	6	9	8	7
6	4	8	7	2	9	1	3	5
7	5	9	1	8	3	6	4	2
9	1	7	2	3	8	5	6	4
8	2	4	6	5	7	3	9	1
5	6	3	9	1	4	7	2	8
3	8	1	4	6	5	2	7	9
4	9	5	3	7	2	8	1	6
2	7	6	8	9	1	4	5	3

Solution # 262

6	7	4	8	3	1	5	9	2
1	3	8	9	5	2	4	6	7
5	2	9	4	6	7	3	8	1
8	4	7	3	2	9	1	5	6
9	6	1	5	7	4	8	2	3
2	5	3	1	8	6	9	7	4
4	1	2	6	9	8	7	3	5
3	9	6	7	1	5	2	4	8
7	8	5	2	4	3	6	1	9

Solution # 263

1	8	5	4	9	3	6	2	7
9	7	3	8	6	2	1	4	5
4	6	2	7	5	1	8	9	3
8	5	4	1	3	6	9	7	2
3	2	1	5	7	9	4	8	6
7	9	6	2	8	4	3	5	1
5	1	9	3	2	8	7	6	4
6	3	7	9	4	5	2	1	8
2	4	8	6	1	7	5	3	9

Solution # 264

2	4	8	7	1	9	5	6	3
3	5	1	6	4	2	9	7	8
7	6	9	3	8	5	1	4	2
5	1	6	2	3	7	8	9	4
9	3	4	8	5	1	6	2	7
8	2	7	4	9	6	3	1	5
6	7	3	9	2	8	4	5	1
4	9	5	1	7	3	2	8	6
1	8	2	5	6	4	7	3	9

Solution # 265

4	9	2	5	1	6	8	3	7
8	3	7	2	9	4	1	6	5
5	1	6	3	8	7	9	4	2
6	7	5	1	2	8	3	9	4
2	8	9	4	3	5	6	7	1
1	4	3	6	7	9	5	2	8
3	2	8	9	4	1	7	5	6
9	5	1	7	6	2	4	8	3
7	6	4	8	5	3	2	1	9

Solution # 266

3	5	1	4	8	7	2	6	9
9	4	6	2	3	5	8	7	1
2	7	8	1	6	9	5	3	4
1	3	5	9	7	4	6	2	8
8	2	4	3	5	6	9	1	7
6	9	7	8	2	1	4	5	3
7	6	9	5	1	8	3	4	2
5	8	3	7	4	2	1	9	6
4	1	2	6	9	3	7	8	5

Solution # 267

8	6	3	1	4	2	7	9	5
9	1	4	5	6	7	2	8	3
5	2	7	8	9	3	4	6	1
4	5	2	7	3	8	9	1	6
6	7	8	9	1	4	3	5	2
1	3	9	6	2	5	8	4	7
7	4	6	3	5	9	1	2	8
2	8	5	4	7	1	6	3	9
3	9	1	2	8	6	5	7	4

Solution # 268

5	9	7	4	3	8	1	2	6
1	3	4	9	6	2	8	7	5
8	2	6	1	5	7	9	4	3
3	7	1	8	4	5	6	9	2
2	4	5	6	7	9	3	1	8
6	8	9	2	1	3	7	5	4
7	6	8	5	9	4	2	3	1
4	1	3	7	2	6	5	8	9
9	5	2	3	8	1	4	6	7

Solution # 269

2	3	8	5	6	9	7	1	4
9	1	6	3	4	7	2	5	8
5	7	4	2	1	8	6	3	9
4	9	2	6	7	3	5	8	1
1	8	7	4	5	2	3	9	6
6	5	3	8	9	1	4	7	2
7	2	5	9	8	6	1	4	3
3	4	9	1	2	5	8	6	7
8	6	1	7	3	4	9	2	5

Solution # 270

3	9	6	8	1	5	4	2	7
8	2	5	3	7	4	9	6	1
4	1	7	2	6	9	8	3	5
6	4	9	1	5	2	3	7	8
1	3	8	9	4	7	2	5	6
5	7	2	6	8	3	1	4	9
2	6	1	7	3	8	5	9	4
9	8	4	5	2	6	7	1	3
7	5	3	4	9	1	6	8	2

Solution # 271

6	1	4	9	5	7	3	8	2
9	8	3	4	6	2	7	5	1
7	2	5	1	3	8	4	9	6
1	9	6	8	4	5	2	3	7
4	5	7	2	9	3	6	1	8
2	3	8	7	1	6	5	4	9
8	7	1	5	2	4	9	6	3
5	6	2	3	8	9	1	7	4
3	4	9	6	7	1	8	2	5

Solution # 272

2	4	3	7	5	9	1	8	6
6	8	9	2	3	1	7	4	5
5	7	1	6	8	4	9	3	2
4	5	6	9	1	3	8	2	7
1	2	8	5	7	6	4	9	3
3	9	7	8	4	2	5	6	1
9	6	4	1	2	7	3	5	8
8	1	2	3	9	5	6	7	4
7	3	5	4	6	8	2	1	9

Solution # 273

3	4	6	7	1	9	8	2	5
7	9	8	2	4	5	6	1	3
5	1	2	3	8	6	7	4	9
8	3	1	6	2	7	9	5	4
4	2	5	1	9	8	3	6	7
6	7	9	4	5	3	2	8	1
2	5	3	8	7	1	4	9	6
9	6	4	5	3	2	1	7	8
1	8	7	9	6	4	5	3	2

Solution # 274

5	1	2	6	7	8	4	9	3
7	3	9	5	4	1	8	6	2
8	6	4	3	9	2	5	1	7
1	8	7	9	3	6	2	5	4
6	2	5	7	1	4	3	8	9
4	9	3	8	2	5	6	7	1
3	5	1	2	6	7	9	4	8
2	7	6	4	8	9	1	3	5
9	4	8	1	5	3	7	2	6

Solution # 275

6	2	3	5	4	1	7	8	9
8	1	9	3	7	6	4	5	2
7	5	4	9	2	8	1	3	6
9	4	6	1	3	7	5	2	8
2	3	7	8	9	5	6	4	1
1	8	5	2	6	4	3	9	7
5	7	1	4	8	2	9	6	3
3	6	2	7	5	9	8	1	4
4	9	8	6	1	3	2	7	5

Solution # 276

1	7	3	2	5	4	9	6	8
6	4	8	9	1	7	3	2	5
5	9	2	8	3	6	7	1	4
8	3	9	4	6	5	2	7	1
7	2	1	3	9	8	5	4	6
4	6	5	1	7	2	8	9	3
2	8	6	5	4	9	1	3	7
9	1	4	7	8	3	6	5	2
3	5	7	6	2	1	4	8	9

Solution # 277

4	9	8	7	3	2	6	1	5
1	2	5	4	8	6	9	3	7
7	6	3	9	1	5	8	2	4
9	3	2	1	7	8	5	4	6
8	5	1	6	9	4	2	7	3
6	4	7	2	5	3	1	8	9
3	7	6	8	2	9	4	5	1
5	8	9	3	4	1	7	6	2
2	1	4	5	6	7	3	9	8

Solution # 278

2	8	3	9	1	7	5	6	4
6	4	7	8	2	5	9	3	1
5	1	9	4	6	3	8	7	2
8	5	4	2	9	6	7	1	3
9	7	6	1	3	4	2	8	5
3	2	1	5	7	8	6	4	9
7	9	2	6	4	1	3	5	8
1	6	5	3	8	2	4	9	7
4	3	8	7	5	9	1	2	6

Solution # 279

5	2	9	4	1	7	3	6	8
6	8	4	5	3	2	7	1	9
7	1	3	9	8	6	5	4	2
1	6	8	3	9	4	2	7	5
9	7	5	8	2	1	4	3	6
4	3	2	7	6	5	8	9	1
2	5	1	6	4	3	9	8	7
8	4	6	2	7	9	1	5	3
3	9	7	1	5	8	6	2	4

Solution # 280

8	5	4	2	9	3	7	6	1
6	2	1	8	7	4	5	9	3
3	7	9	1	6	5	8	2	4
2	1	5	6	3	7	9	4	8
9	3	8	5	4	1	6	7	2
7	4	6	9	2	8	1	3	5
5	9	3	7	1	2	4	8	6
1	6	2	4	8	9	3	5	7
4	8	7	3	5	6	2	1	9

Solution # 281

6	2	1	7	5	9	3	8	4
7	4	3	6	8	1	9	2	5
9	5	8	2	4	3	7	6	1
2	3	9	1	6	8	4	5	7
5	1	7	3	2	4	8	9	6
4	8	6	9	7	5	2	1	3
8	9	4	5	3	6	1	7	2
3	7	5	8	1	2	6	4	9
1	6	2	4	9	7	5	3	8

Solution # 282

9	5	4	7	2	1	8	6	3
3	7	1	8	5	6	2	9	4
8	2	6	9	4	3	7	5	1
2	4	9	6	1	8	5	3	7
6	3	8	2	7	5	4	1	9
5	1	7	3	9	4	6	2	8
7	8	2	5	3	9	1	4	6
1	9	5	4	6	7	3	8	2
4	6	3	1	8	2	9	7	5

Solution # 283

5	3	7	6	1	4	2	9	8
4	8	9	5	3	2	1	7	6
1	2	6	8	9	7	4	5	3
9	6	2	1	8	5	3	4	7
7	1	4	9	2	3	6	8	5
8	5	3	7	4	6	9	2	1
3	9	8	2	5	1	7	6	4
2	7	1	4	6	8	5	3	9
6	4	5	3	7	9	8	1	2

Solution # 284

3	4	7	5	9	8	1	6	2
2	9	6	3	7	1	8	4	5
1	8	5	2	6	4	3	7	9
6	7	1	9	8	5	4	2	3
9	3	8	6	4	2	7	5	1
4	5	2	7	1	3	6	9	8
5	1	4	8	2	7	9	3	6
7	2	9	1	3	6	5	8	4
8	6	3	4	5	9	2	1	7

Solution # 285

5	3	4	6	1	9	2	8	7
9	7	8	2	3	4	6	5	1
6	2	1	5	8	7	4	9	3
7	1	5	9	4	6	3	2	8
3	8	9	1	7	2	5	6	4
2	4	6	3	5	8	7	1	9
8	9	3	7	6	5	1	4	2
4	6	7	8	2	1	9	3	5
1	5	2	4	9	3	8	7	6

Solution # 286

9	8	2	7	1	3	4	6	5
5	4	1	6	9	8	7	3	2
3	7	6	2	5	4	1	9	8
8	5	9	1	6	2	3	7	4
4	6	7	9	3	5	2	8	1
1	2	3	4	8	7	9	5	6
6	1	4	5	7	9	8	2	3
7	3	5	8	2	1	6	4	9
2	9	8	3	4	6	5	1	7

Solution # 287

4	3	9	8	7	5	1	2	6
7	1	5	9	6	2	8	3	4
8	6	2	3	1	4	7	5	9
2	9	4	6	5	1	3	7	8
1	8	7	2	4	3	6	9	5
3	5	6	7	9	8	4	1	2
6	7	8	5	3	9	2	4	1
5	4	3	1	2	6	9	8	7
9	2	1	4	8	7	5	6	3

Solution # 288

7	9	8	5	3	6	1	2	4
1	6	4	9	2	7	3	5	8
3	2	5	4	8	1	7	9	6
6	1	9	7	4	2	8	3	5
2	8	3	6	5	9	4	7	1
5	4	7	8	1	3	2	6	9
8	5	6	3	7	4	9	1	2
9	3	2	1	6	8	5	4	7
4	7	1	2	9	5	6	8	3

Solution # 289

4	7	6	9	3	2	1	8	5
5	2	9	1	8	4	6	7	3
8	3	1	5	7	6	9	4	2
7	6	3	4	9	8	2	5	1
9	5	4	7	2	1	8	3	6
1	8	2	6	5	3	7	9	4
2	9	7	3	1	5	4	6	8
6	1	5	8	4	7	3	2	9
3	4	8	2	6	9	5	1	7

Solution # 290

3	9	4	6	5	8	2	1	7
7	8	2	1	3	9	4	6	5
6	5	1	2	4	7	3	9	8
5	2	6	3	8	1	7	4	9
9	1	7	5	2	4	6	8	3
4	3	8	7	9	6	1	5	2
8	7	9	4	6	3	5	2	1
1	4	5	8	7	2	9	3	6
2	6	3	9	1	5	8	7	4

Solution # 291

7	8	9	5	3	4	1	2	6
4	1	2	6	8	7	9	3	5
6	3	5	9	1	2	8	7	4
3	5	6	7	9	1	4	8	2
9	4	8	2	5	3	7	6	1
1	2	7	4	6	8	5	9	3
5	7	4	3	2	9	6	1	8
8	6	3	1	7	5	2	4	9
2	9	1	8	4	6	3	5	7

Solution # 292

9	3	5	6	8	1	7	2	4
7	8	2	5	3	4	6	1	9
6	4	1	7	9	2	8	5	3
2	9	3	8	6	5	1	4	7
4	5	7	1	2	3	9	6	8
8	1	6	4	7	9	5	3	2
1	7	9	2	4	6	3	8	5
5	2	8	3	1	7	4	9	6
3	6	4	9	5	8	2	7	1

Solution # 293

5	3	1	9	7	6	8	2	4
9	2	7	1	4	8	3	5	6
8	6	4	5	3	2	1	7	9
6	1	5	3	2	9	7	4	8
2	4	9	7	8	5	6	3	1
3	7	8	4	6	1	2	9	5
4	8	6	2	9	3	5	1	7
7	5	2	6	1	4	9	8	3
1	9	3	8	5	7	4	6	2

Solution # 294

2	6	7	5	3	9	4	8	1
4	3	8	7	1	2	6	5	9
9	5	1	6	4	8	3	2	7
5	1	2	4	9	3	7	6	8
3	4	6	1	8	7	2	9	5
8	7	9	2	5	6	1	3	4
7	8	5	3	6	4	9	1	2
6	9	4	8	2	1	5	7	3
1	2	3	9	7	5	8	4	6

Solution # 295

4	8	1	7	6	3	2	9	5
9	7	3	2	8	5	4	6	1
6	2	5	4	9	1	3	7	8
2	4	8	9	1	7	5	3	6
1	9	7	3	5	6	8	4	2
5	3	6	8	4	2	9	1	7
8	6	2	1	3	9	7	5	4
3	5	4	6	7	8	1	2	9
7	1	9	5	2	4	6	8	3

Solution # 296

4	3	5	7	1	2	8	6	9
2	8	9	5	6	4	3	7	1
6	7	1	9	3	8	5	2	4
5	1	3	8	7	6	4	9	2
7	2	8	1	4	9	6	3	5
9	4	6	3	2	5	7	1	8
3	5	7	2	8	1	9	4	6
8	6	2	4	9	3	1	5	7
1	9	4	6	5	7	2	8	3

Solution # 297

7	9	8	4	3	6	5	1	2
5	3	4	1	9	2	8	6	7
6	1	2	8	5	7	9	3	4
3	4	6	9	1	5	7	2	8
2	8	1	7	6	4	3	5	9
9	5	7	3	2	8	1	4	6
4	2	9	5	7	1	6	8	3
8	7	5	6	4	3	2	9	1
1	6	3	2	8	9	4	7	5

Solution # 298

3	2	7	6	1	5	9	4	8
1	6	8	9	4	7	5	3	2
9	5	4	8	3	2	1	6	7
4	1	5	2	6	8	3	7	9
6	7	2	5	9	3	4	8	1
8	9	3	4	7	1	6	2	5
2	4	9	1	8	6	7	5	3
5	3	6	7	2	9	8	1	4
7	8	1	3	5	4	2	9	6

Solution # 299

7	6	2	5	8	4	1	9	3
1	8	5	6	9	3	2	7	4
4	9	3	2	7	1	8	5	6
8	5	6	4	2	9	3	1	7
3	2	7	1	5	8	6	4	9
9	1	4	7	3	6	5	8	2
6	3	1	8	4	7	9	2	5
5	4	9	3	1	2	7	6	8
2	7	8	9	6	5	4	3	1

Solution # 300

1	8	2	3	6	7	5	9	4
6	3	5	4	8	9	2	1	7
7	4	9	5	2	1	3	8	6
8	1	3	7	9	2	4	6	5
5	7	6	1	4	8	9	2	3
9	2	4	6	5	3	8	7	1
3	6	8	2	7	5	1	4	9
4	9	1	8	3	6	7	5	2
2	5	7	9	1	4	6	3	8

Solution # 301

6	5	9	1	4	2	8	7	3
4	8	3	9	7	6	1	5	2
7	2	1	3	8	5	9	6	4
1	9	2	7	5	4	6	3	8
5	7	8	6	9	3	2	4	1
3	4	6	2	1	8	7	9	5
2	6	4	8	3	9	5	1	7
8	3	7	5	6	1	4	2	9
9	1	5	4	2	7	3	8	6

Solution # 302

9	3	7	6	1	4	2	5	8
8	6	1	2	5	7	4	3	9
2	5	4	3	9	8	7	6	1
5	8	3	7	4	1	9	2	6
6	4	9	5	2	3	8	1	7
7	1	2	9	8	6	5	4	3
3	7	8	4	6	2	1	9	5
1	2	5	8	3	9	6	7	4
4	9	6	1	7	5	3	8	2

Solution # 303

9	5	1	4	8	6	7	2	3
7	3	4	1	5	2	8	9	6
2	8	6	7	3	9	4	1	5
8	2	5	3	1	4	6	7	9
6	9	3	2	7	8	1	5	4
1	4	7	9	6	5	3	8	2
4	1	9	6	2	7	5	3	8
3	6	8	5	9	1	2	4	7
5	7	2	8	4	3	9	6	1

Solution # 304

3	9	7	2	4	6	5	1	8
5	2	8	7	1	9	4	3	6
1	4	6	3	8	5	2	9	7
4	6	9	5	7	2	3	8	1
2	7	1	6	3	8	9	4	5
8	5	3	4	9	1	6	7	2
7	8	5	9	2	4	1	6	3
9	1	2	8	6	3	7	5	4
6	3	4	1	5	7	8	2	9

Solution # 305

5	9	8	6	1	7	3	2	4
7	6	2	3	4	8	1	5	9
3	1	4	5	9	2	8	7	6
2	7	9	8	6	1	5	4	3
8	4	6	9	5	3	2	1	7
1	3	5	2	7	4	9	6	8
9	2	1	4	8	6	7	3	5
6	5	3	7	2	9	4	8	1
4	8	7	1	3	5	6	9	2

Solution # 306

6	8	1	4	9	7	5	2	3
2	4	3	6	5	8	7	9	1
7	9	5	3	2	1	8	6	4
3	2	7	9	6	4	1	8	5
4	6	9	8	1	5	2	3	7
1	5	8	7	3	2	9	4	6
9	7	6	5	8	3	4	1	2
5	3	2	1	4	9	6	7	8
8	1	4	2	7	6	3	5	9

Solution # 307

4	8	7	9	3	1	6	5	2
5	9	1	2	6	7	8	3	4
2	3	6	8	5	4	7	9	1
6	5	9	1	4	2	3	7	8
1	7	2	3	8	5	4	6	9
8	4	3	7	9	6	2	1	5
3	6	8	5	2	9	1	4	7
7	2	5	4	1	3	9	8	6
9	1	4	6	7	8	5	2	3

Solution # 308

3	5	9	8	2	4	7	6	1
7	8	1	6	3	9	4	5	2
6	2	4	7	5	1	9	3	8
8	3	7	9	4	2	5	1	6
4	1	5	3	8	6	2	7	9
9	6	2	5	1	7	3	8	4
2	7	6	1	9	5	8	4	3
5	9	8	4	6	3	1	2	7
1	4	3	2	7	8	6	9	5

Solution # 309

7	1	5	2	9	8	3	4	6
6	9	8	3	1	4	7	2	5
2	3	4	7	5	6	8	1	9
1	5	2	4	7	3	6	9	8
3	4	6	9	8	5	1	7	2
8	7	9	6	2	1	4	5	3
9	6	7	8	4	2	5	3	1
4	8	1	5	3	9	2	6	7
5	2	3	1	6	7	9	8	4

Solution # 310

1	4	5	8	7	3	2	9	6
6	3	8	9	2	1	7	5	4
2	9	7	5	6	4	8	1	3
5	8	4	6	3	9	1	2	7
3	2	1	7	5	8	4	6	9
9	7	6	4	1	2	5	3	8
7	5	9	1	8	6	3	4	2
8	6	3	2	4	5	9	7	1
4	1	2	3	9	7	6	8	5

Solution # 311

9	5	1	4	3	2	6	7	8
2	3	7	8	1	6	4	5	9
6	8	4	5	9	7	1	3	2
3	4	5	7	8	9	2	1	6
8	7	9	6	2	1	3	4	5
1	2	6	3	5	4	8	9	7
7	6	8	9	4	3	5	2	1
5	1	3	2	7	8	9	6	4
4	9	2	1	6	5	7	8	3

Solution # 312

1	6	4	9	7	8	2	3	5
2	3	9	1	6	5	8	4	7
8	7	5	2	3	4	6	9	1
5	1	6	3	8	9	4	7	2
4	9	7	5	2	6	3	1	8
3	8	2	7	4	1	5	6	9
7	4	1	8	5	3	9	2	6
9	5	3	6	1	2	7	8	4
6	2	8	4	9	7	1	5	3

Solution # 313

7	9	3	5	2	6	4	8	1
2	5	1	4	8	7	9	3	6
6	4	8	3	1	9	7	2	5
8	7	9	6	3	1	2	5	4
3	6	2	8	4	5	1	9	7
4	1	5	7	9	2	3	6	8
9	2	6	1	5	4	8	7	3
1	8	7	2	6	3	5	4	9
5	3	4	9	7	8	6	1	2

Solution # 314

3	6	9	4	5	7	2	1	8
7	8	1	2	9	6	4	3	5
5	2	4	1	8	3	7	9	6
9	1	2	3	6	5	8	4	7
8	5	6	7	4	9	1	2	3
4	7	3	8	1	2	5	6	9
6	9	7	5	2	4	3	8	1
2	3	8	6	7	1	9	5	4
1	4	5	9	3	8	6	7	2

Solution # 315

1	7	5	9	6	3	4	2	8
9	2	6	4	1	8	3	7	5
4	3	8	7	2	5	1	9	6
7	8	3	5	4	2	9	6	1
6	1	2	3	9	7	8	5	4
5	4	9	1	8	6	7	3	2
2	6	7	8	3	1	5	4	9
3	9	1	6	5	4	2	8	7
8	5	4	2	7	9	6	1	3

Solution # 316

1	5	8	9	6	4	2	3	7
4	9	6	2	7	3	5	8	1
2	3	7	5	8	1	4	6	9
9	6	3	7	4	8	1	2	5
8	2	5	1	3	9	6	7	4
7	4	1	6	2	5	3	9	8
6	8	4	3	1	7	9	5	2
5	1	2	8	9	6	7	4	3
3	7	9	4	5	2	8	1	6

Solution # 317

6	8	3	2	5	4	9	1	7
9	5	7	1	6	8	2	4	3
2	4	1	9	3	7	6	5	8
3	6	8	5	7	2	1	9	4
7	2	9	4	8	1	5	3	6
5	1	4	6	9	3	8	7	2
4	9	5	7	2	6	3	8	1
1	3	6	8	4	5	7	2	9
8	7	2	3	1	9	4	6	5

Solution # 318

5	1	4	6	7	2	9	3	8
6	3	8	5	9	1	4	2	7
2	9	7	4	8	3	5	6	1
4	2	3	1	5	9	7	8	6
7	6	1	3	4	8	2	9	5
8	5	9	2	6	7	1	4	3
9	8	2	7	3	5	6	1	4
3	7	6	9	1	4	8	5	2
1	4	5	8	2	6	3	7	9

Solution # 319

9	4	2	8	6	7	3	1	5
5	8	3	2	9	1	7	6	4
7	1	6	3	5	4	2	9	8
1	9	8	6	2	5	4	3	7
6	7	5	4	8	3	9	2	1
2	3	4	1	7	9	8	5	6
8	5	7	9	3	6	1	4	2
3	6	1	7	4	2	5	8	9
4	2	9	5	1	8	6	7	3

Solution # 320

6	8	2	4	3	9	1	5	7
5	4	7	6	1	2	9	8	3
9	1	3	8	5	7	4	2	6
4	3	5	1	2	6	7	9	8
7	6	1	5	9	8	2	3	4
8	2	9	3	7	4	6	1	5
2	9	8	7	4	5	3	6	1
3	7	6	9	8	1	5	4	2
1	5	4	2	6	3	8	7	9

Solution # 321

2	9	4	7	8	6	3	1	5
7	8	3	5	4	1	6	9	2
1	6	5	9	3	2	7	8	4
4	2	8	3	1	5	9	6	7
6	3	7	2	9	8	5	4	1
9	5	1	4	6	7	2	3	8
5	1	6	8	2	3	4	7	9
8	7	9	6	5	4	1	2	3
3	4	2	1	7	9	8	5	6

Solution # 322

2	6	3	4	8	7	1	9	5
4	7	5	9	2	1	3	8	6
1	8	9	5	3	6	7	4	2
5	9	6	1	7	2	8	3	4
3	4	7	8	6	9	2	5	1
8	1	2	3	5	4	9	6	7
7	2	8	6	9	5	4	1	3
6	3	4	7	1	8	5	2	9
9	5	1	2	4	3	6	7	8

Solution # 323

5	6	7	9	3	1	2	4	8
1	9	2	4	5	8	6	7	3
4	8	3	7	2	6	5	9	1
9	4	6	5	8	2	1	3	7
3	1	5	6	9	7	4	8	2
2	7	8	1	4	3	9	6	5
8	5	9	3	1	4	7	2	6
7	3	4	2	6	5	8	1	9
6	2	1	8	7	9	3	5	4

Solution # 324

6	2	8	3	9	4	7	1	5
7	9	1	2	8	5	3	6	4
3	4	5	1	6	7	9	2	8
2	7	3	4	5	9	6	8	1
8	5	9	6	1	3	4	7	2
1	6	4	8	7	2	5	9	3
4	3	6	7	2	8	1	5	9
5	1	2	9	4	6	8	3	7
9	8	7	5	3	1	2	4	6

Solution # 325

8	5	1	9	4	6	3	2	7
2	9	6	3	7	1	8	4	5
7	4	3	8	2	5	1	6	9
4	3	2	5	9	8	7	1	6
6	1	5	2	3	7	9	8	4
9	8	7	1	6	4	2	5	3
5	7	8	6	1	9	4	3	2
1	2	4	7	5	3	6	9	8
3	6	9	4	8	2	5	7	1

Solution # 326

2	6	1	5	7	9	8	4	3
9	5	8	2	4	3	6	7	1
3	4	7	1	8	6	5	9	2
4	1	2	9	6	5	3	8	7
6	8	9	4	3	7	2	1	5
7	3	5	8	1	2	9	6	4
5	2	4	7	9	8	1	3	6
1	9	6	3	5	4	7	2	8
8	7	3	6	2	1	4	5	9

Solution # 327

8	3	2	7	4	6	1	9	5
4	7	5	8	9	1	3	6	2
6	9	1	2	3	5	4	8	7
1	6	8	5	7	3	2	4	9
9	4	7	6	2	8	5	3	1
5	2	3	9	1	4	6	7	8
7	1	9	3	6	2	8	5	4
3	8	4	1	5	9	7	2	6
2	5	6	4	8	7	9	1	3

Solution # 328

2	8	7	6	9	3	5	4	1
9	3	4	2	5	1	6	8	7
1	5	6	7	8	4	2	9	3
7	2	5	8	1	6	9	3	4
4	6	8	9	3	2	7	1	5
3	9	1	4	7	5	8	2	6
5	4	9	1	2	7	3	6	8
8	1	3	5	6	9	4	7	2
6	7	2	3	4	8	1	5	9

Solution # 329

8	9	5	7	4	6	3	2	1
3	6	4	2	5	1	9	7	8
2	7	1	3	8	9	4	5	6
1	5	7	6	9	4	8	3	2
6	4	2	1	3	8	7	9	5
9	8	3	5	2	7	1	6	4
5	1	8	9	6	3	2	4	7
4	2	9	8	7	5	6	1	3
7	3	6	4	1	2	5	8	9

Solution # 330

6	8	1	5	9	4	2	3	7
5	4	7	2	8	3	1	9	6
2	3	9	1	6	7	4	5	8
3	5	6	7	2	8	9	4	1
8	9	2	4	5	1	6	7	3
1	7	4	9	3	6	5	8	2
4	6	5	3	7	2	8	1	9
9	2	3	8	1	5	7	6	4
7	1	8	6	4	9	3	2	5

Solution # 331

9	5	7	3	1	8	4	2	6
6	3	1	4	2	7	8	5	9
4	2	8	5	9	6	7	3	1
2	6	5	7	8	1	9	4	3
3	8	9	2	4	5	6	1	7
7	1	4	9	6	3	2	8	5
5	7	6	8	3	4	1	9	2
1	4	2	6	5	9	3	7	8
8	9	3	1	7	2	5	6	4

Solution # 332

7	2	8	5	9	1	4	3	6
9	1	5	3	6	4	8	2	7
4	3	6	7	2	8	5	1	9
1	9	3	2	7	5	6	4	8
8	6	4	1	3	9	7	5	2
2	5	7	8	4	6	1	9	3
3	8	1	6	5	2	9	7	4
6	4	2	9	1	7	3	8	5
5	7	9	4	8	3	2	6	1

Solution # 333

3	2	4	6	8	9	1	5	7
6	5	8	4	1	7	3	2	9
1	7	9	2	5	3	6	4	8
9	3	7	1	6	5	4	8	2
4	1	2	7	9	8	5	6	3
8	6	5	3	4	2	7	9	1
7	8	3	5	2	4	9	1	6
5	9	1	8	7	6	2	3	4
2	4	6	9	3	1	8	7	5

Solution # 334

5	2	6	9	3	7	1	4	8
7	4	3	8	2	1	5	9	6
9	8	1	5	4	6	2	3	7
3	9	8	1	6	4	7	5	2
2	1	5	7	9	8	3	6	4
4	6	7	2	5	3	8	1	9
6	5	4	3	8	2	9	7	1
8	7	9	6	1	5	4	2	3
1	3	2	4	7	9	6	8	5

Solution # 335

1	6	4	5	2	3	9	7	8
7	8	5	1	9	4	3	2	6
9	3	2	7	8	6	5	1	4
3	1	8	2	7	5	4	6	9
5	7	9	4	6	1	8	3	2
4	2	6	8	3	9	1	5	7
2	5	7	3	4	8	6	9	1
6	4	1	9	5	2	7	8	3
8	9	3	6	1	7	2	4	5

Solution # 336

1	6	4	8	2	5	9	7	3
2	3	9	6	4	7	1	5	8
8	5	7	1	3	9	6	4	2
6	4	8	3	7	1	2	9	5
5	9	1	4	8	2	7	3	6
7	2	3	5	9	6	4	8	1
4	1	2	7	5	8	3	6	9
3	8	6	9	1	4	5	2	7
9	7	5	2	6	3	8	1	4

Solution # 337

3	2	8	6	7	1	9	5	4
4	7	9	8	2	5	1	3	6
5	6	1	4	3	9	2	7	8
1	9	3	7	6	8	5	4	2
6	8	2	5	4	3	7	9	1
7	5	4	9	1	2	8	6	3
2	1	6	3	5	7	4	8	9
9	4	5	1	8	6	3	2	7
8	3	7	2	9	4	6	1	5

Solution # 338

9	6	3	5	4	8	1	7	2
5	8	1	7	9	2	6	4	3
7	2	4	6	3	1	8	5	9
2	5	7	9	8	3	4	6	1
4	3	9	1	5	6	2	8	7
6	1	8	2	7	4	9	3	5
8	7	5	4	1	9	3	2	6
1	4	6	3	2	7	5	9	8
3	9	2	8	6	5	7	1	4

Solution # 339

3	5	6	7	8	4	1	2	9
4	2	7	1	5	9	8	3	6
1	9	8	6	2	3	4	5	7
8	3	5	2	1	7	6	9	4
2	1	9	5	4	6	7	8	3
7	6	4	9	3	8	5	1	2
6	7	2	8	9	1	3	4	5
9	8	3	4	7	5	2	6	1
5	4	1	3	6	2	9	7	8

Solution # 340

3	7	1	2	9	6	4	5	8
9	5	4	3	7	8	6	1	2
6	2	8	5	4	1	7	3	9
5	4	7	1	3	9	2	8	6
2	6	3	8	5	4	1	9	7
1	8	9	6	2	7	5	4	3
4	3	2	9	6	5	8	7	1
8	9	5	7	1	2	3	6	4
7	1	6	4	8	3	9	2	5

Solution # 341

7	3	4	9	1	2	8	6	5
9	8	2	6	4	5	3	1	7
5	1	6	7	8	3	4	9	2
1	5	9	3	6	7	2	4	8
4	2	7	8	9	1	6	5	3
3	6	8	5	2	4	1	7	9
6	9	3	4	7	8	5	2	1
8	7	1	2	5	6	9	3	4
2	4	5	1	3	9	7	8	6

Solution # 342

4	5	8	6	1	3	2	9	7
3	9	6	2	5	7	4	8	1
2	7	1	4	8	9	3	5	6
6	4	5	8	9	1	7	3	2
7	1	9	5	3	2	8	6	4
8	2	3	7	6	4	9	1	5
9	3	7	1	2	6	5	4	8
5	6	4	9	7	8	1	2	3
1	8	2	3	4	5	6	7	9

Solution # 343

3	1	7	6	8	9	5	4	2
8	4	6	3	2	5	9	1	7
5	2	9	4	7	1	8	3	6
4	6	1	8	5	3	2	7	9
2	9	8	7	6	4	1	5	3
7	3	5	9	1	2	4	6	8
9	8	4	5	3	6	7	2	1
6	7	2	1	4	8	3	9	5
1	5	3	2	9	7	6	8	4

Solution # 344

9	6	8	4	3	1	5	2	7
4	7	1	2	5	6	9	3	8
3	5	2	8	7	9	6	4	1
8	9	7	3	4	2	1	5	6
6	3	5	1	9	7	4	8	2
1	2	4	5	6	8	7	9	3
5	1	6	9	8	3	2	7	4
2	4	3	7	1	5	8	6	9
7	8	9	6	2	4	3	1	5

Solution # 345

1	6	9	8	4	7	5	2	3
7	4	2	5	6	3	9	8	1
3	5	8	9	1	2	4	6	7
6	8	3	7	5	4	1	9	2
4	9	7	2	8	1	6	3	5
5	2	1	6	3	9	8	7	4
2	7	5	1	9	8	3	4	6
8	1	4	3	7	6	2	5	9
9	3	6	4	2	5	7	1	8

Solution # 346

3	8	1	4	2	7	9	5	6
2	9	5	6	8	1	7	3	4
7	4	6	9	3	5	2	8	1
5	3	9	7	1	4	8	6	2
4	6	2	3	9	8	1	7	5
8	1	7	5	6	2	3	4	9
1	5	4	2	7	3	6	9	8
9	2	3	8	4	6	5	1	7
6	7	8	1	5	9	4	2	3

Solution # 347

8	2	3	5	4	1	6	9	7
4	7	5	9	3	6	1	2	8
6	1	9	8	2	7	4	3	5
5	4	7	3	9	2	8	6	1
3	6	1	4	7	8	9	5	2
2	9	8	1	6	5	7	4	3
7	3	4	2	1	9	5	8	6
9	8	6	7	5	3	2	1	4
1	5	2	6	8	4	3	7	9

Solution # 348

6	3	9	7	8	1	5	4	2
2	1	8	4	5	3	7	6	9
7	5	4	2	9	6	3	1	8
5	8	6	1	7	9	2	3	4
3	7	1	6	2	4	8	9	5
4	9	2	5	3	8	6	7	1
1	2	5	3	4	7	9	8	6
9	6	7	8	1	2	4	5	3
8	4	3	9	6	5	1	2	7

Solution # 349

2	9	3	5	1	6	4	7	8
6	1	4	7	3	8	9	2	5
5	8	7	2	4	9	3	1	6
7	2	9	4	8	1	5	6	3
4	6	1	3	9	5	2	8	7
3	5	8	6	7	2	1	9	4
1	4	6	8	2	3	7	5	9
9	3	5	1	6	7	8	4	2
8	7	2	9	5	4	6	3	1

Solution # 350

9	5	3	6	7	1	2	8	4
1	4	8	5	9	2	3	6	7
6	7	2	4	8	3	1	9	5
4	6	7	8	3	9	5	2	1
8	3	1	2	5	4	9	7	6
2	9	5	7	1	6	8	4	3
5	8	6	1	2	7	4	3	9
7	1	9	3	4	8	6	5	2
3	2	4	9	6	5	7	1	8

Solution # 351

6	7	5	4	8	3	1	2	9
4	8	9	2	7	1	6	3	5
2	1	3	5	6	9	4	7	8
5	6	1	3	2	7	8	9	4
8	3	2	9	5	4	7	6	1
7	9	4	6	1	8	2	5	3
3	5	7	8	4	2	9	1	6
9	2	8	1	3	6	5	4	7
1	4	6	7	9	5	3	8	2

Solution # 352

2	3	4	9	1	6	7	8	5
5	6	9	7	8	3	4	1	2
7	8	1	2	4	5	3	6	9
4	5	8	1	9	7	2	3	6
9	2	7	6	3	8	1	5	4
3	1	6	4	5	2	9	7	8
8	7	2	3	6	4	5	9	1
6	9	3	5	2	1	8	4	7
1	4	5	8	7	9	6	2	3

Solution # 353

2	7	6	1	4	9	3	5	8
4	1	8	5	3	6	7	9	2
9	5	3	8	7	2	6	1	4
1	3	2	6	8	4	5	7	9
7	6	9	2	1	5	8	4	3
5	8	4	7	9	3	2	6	1
8	4	1	3	5	7	9	2	6
3	2	5	9	6	1	4	8	7
6	9	7	4	2	8	1	3	5

Solution # 354

1	2	7	6	9	3	8	4	5
8	3	6	5	2	4	9	1	7
5	9	4	1	7	8	3	2	6
4	5	8	2	3	6	1	7	9
3	6	1	9	5	7	2	8	4
2	7	9	4	8	1	5	6	3
9	1	5	7	6	2	4	3	8
7	4	3	8	1	5	6	9	2
6	8	2	3	4	9	7	5	1

Solution # 355

5	7	3	6	1	2	8	4	9
2	4	8	5	9	7	1	6	3
6	9	1	4	3	8	5	7	2
1	5	9	2	6	3	4	8	7
4	2	7	8	5	1	9	3	6
8	3	6	9	7	4	2	1	5
3	6	4	1	2	5	7	9	8
9	1	2	7	8	6	3	5	4
7	8	5	3	4	9	6	2	1

Solution # 356

1	5	6	4	2	9	8	3	7
8	4	9	3	1	7	2	6	5
7	2	3	6	5	8	4	1	9
4	1	2	7	6	3	9	5	8
6	9	5	8	4	2	1	7	3
3	8	7	1	9	5	6	4	2
5	6	8	9	3	1	7	2	4
9	3	4	2	7	6	5	8	1
2	7	1	5	8	4	3	9	6

Solution # 357

4	6	5	2	9	8	7	1	3
2	7	8	3	4	1	5	6	9
1	9	3	7	5	6	4	2	8
8	1	7	6	3	9	2	4	5
3	5	2	4	8	7	1	9	6
9	4	6	1	2	5	8	3	7
6	8	4	9	7	2	3	5	1
5	3	1	8	6	4	9	7	2
7	2	9	5	1	3	6	8	4

Solution # 358

2	4	1	7	3	8	5	9	6
8	6	9	2	5	4	3	1	7
3	5	7	9	1	6	4	2	8
6	9	2	3	8	1	7	5	4
4	7	3	5	2	9	8	6	1
5	1	8	4	6	7	9	3	2
9	2	6	8	7	3	1	4	5
7	3	5	1	4	2	6	8	9
1	8	4	6	9	5	2	7	3

Solution # 359

5	6	1	9	7	3	2	8	4
4	9	7	2	8	6	3	5	1
8	3	2	5	1	4	9	6	7
9	4	5	3	2	1	6	7	8
1	7	8	6	5	9	4	2	3
3	2	6	8	4	7	1	9	5
7	8	3	4	6	2	5	1	9
2	1	9	7	3	5	8	4	6
6	5	4	1	9	8	7	3	2

Solution # 360

8	6	5	3	1	4	9	7	2
4	1	2	7	9	5	6	8	3
3	9	7	6	2	8	4	5	1
9	4	1	8	5	3	7	2	6
7	3	8	2	4	6	5	1	9
2	5	6	1	7	9	3	4	8
6	2	4	9	8	7	1	3	5
5	8	3	4	6	1	2	9	7
1	7	9	5	3	2	8	6	4

Solution # 361

2	3	1	9	4	5	7	8	6
9	8	4	7	6	3	2	5	1
5	7	6	1	2	8	3	9	4
4	9	3	6	5	2	1	7	8
7	5	2	4	8	1	6	3	9
6	1	8	3	7	9	5	4	2
3	2	7	8	1	4	9	6	5
8	6	5	2	9	7	4	1	3
1	4	9	5	3	6	8	2	7

Solution # 362

4	9	5	7	2	3	1	6	8
1	2	6	4	8	9	7	3	5
8	3	7	1	6	5	4	2	9
7	6	3	9	5	4	8	1	2
9	5	1	8	7	2	3	4	6
2	4	8	3	1	6	5	9	7
3	8	9	6	4	7	2	5	1
6	1	2	5	3	8	9	7	4
5	7	4	2	9	1	6	8	3

Solution # 363

9	3	4	1	6	7	2	5	8
2	5	1	8	4	3	7	9	6
6	8	7	2	5	9	4	3	1
7	4	5	6	3	2	1	8	9
8	6	2	7	9	1	3	4	5
1	9	3	4	8	5	6	7	2
3	2	9	5	7	6	8	1	4
4	7	6	9	1	8	5	2	3
5	1	8	3	2	4	9	6	7

Solution # 364

3	1	5	4	7	6	2	8	9
2	8	6	5	9	3	4	7	1
7	4	9	8	2	1	5	6	3
8	2	4	9	1	7	6	3	5
1	6	3	2	4	5	8	9	7
9	5	7	3	6	8	1	2	4
5	7	1	6	8	9	3	4	2
4	9	8	1	3	2	7	5	6
6	3	2	7	5	4	9	1	8

Solution # 365

9	8	3	6	4	5	2	7	1
1	5	2	8	7	3	4	6	9
4	6	7	2	1	9	3	8	5
2	9	4	3	6	1	8	5	7
8	7	6	5	2	4	1	9	3
5	3	1	7	9	8	6	4	2
3	2	5	4	8	7	9	1	6
7	4	9	1	3	6	5	2	8
6	1	8	9	5	2	7	3	4

Solution # 366

7	5	1	4	9	8	2	3	6
4	9	2	6	5	3	8	1	7
3	8	6	1	2	7	5	4	9
1	6	7	5	4	2	9	8	3
5	4	9	3	8	1	7	6	2
8	2	3	7	6	9	1	5	4
2	3	4	8	7	5	6	9	1
6	7	8	9	1	4	3	2	5
9	1	5	2	3	6	4	7	8

Solution # 367

1	5	4	8	9	2	6	3	7
6	2	8	7	1	3	4	5	9
3	7	9	5	6	4	8	1	2
4	6	1	3	5	9	7	2	8
8	3	2	1	7	6	5	9	4
5	9	7	4	2	8	1	6	3
7	8	5	2	3	1	9	4	6
9	4	3	6	8	5	2	7	1
2	1	6	9	4	7	3	8	5

Solution # 368

8	1	3	2	4	7	6	5	9
5	2	4	9	3	6	8	7	1
9	6	7	1	5	8	4	3	2
1	7	9	6	2	4	5	8	3
2	8	5	7	9	3	1	4	6
4	3	6	5	8	1	2	9	7
7	4	8	3	1	2	9	6	5
6	9	1	4	7	5	3	2	8
3	5	2	8	6	9	7	1	4

Solution # 369

6	5	1	8	2	4	3	7	9
7	8	4	1	9	3	2	6	5
9	2	3	6	5	7	1	4	8
2	6	7	5	1	8	4	9	3
8	4	5	2	3	9	6	1	7
3	1	9	4	7	6	5	8	2
5	7	2	9	6	1	8	3	4
4	9	6	3	8	2	7	5	1
1	3	8	7	4	5	9	2	6

Solution # 370

7	8	4	5	9	2	1	6	3
3	6	9	1	8	7	2	5	4
1	5	2	4	3	6	9	7	8
5	3	8	9	6	4	7	2	1
6	9	7	2	1	3	4	8	5
2	4	1	8	7	5	6	3	9
8	7	6	3	4	1	5	9	2
4	2	3	6	5	9	8	1	7
9	1	5	7	2	8	3	4	6

Solution # 371

8	6	7	4	3	2	1	5	9
1	4	2	6	5	9	8	3	7
5	3	9	7	8	1	2	4	6
4	2	3	9	7	6	5	1	8
7	9	8	5	1	4	6	2	3
6	5	1	3	2	8	9	7	4
3	8	4	1	9	5	7	6	2
2	7	5	8	6	3	4	9	1
9	1	6	2	4	7	3	8	5

Solution # 372

1	9	4	7	2	6	8	3	5
7	3	5	8	9	4	1	6	2
6	2	8	3	1	5	4	9	7
4	1	2	5	3	8	9	7	6
5	7	3	4	6	9	2	1	8
8	6	9	2	7	1	3	5	4
9	4	7	1	5	2	6	8	3
3	8	6	9	4	7	5	2	1
2	5	1	6	8	3	7	4	9

Solution # 373

1	9	6	7	2	3	4	8	5
7	8	2	5	4	9	1	6	3
4	3	5	6	8	1	2	7	9
2	5	9	1	3	7	8	4	6
8	1	4	2	5	6	9	3	7
3	6	7	4	9	8	5	1	2
5	2	1	3	6	4	7	9	8
6	7	8	9	1	5	3	2	4
9	4	3	8	7	2	6	5	1

Solution # 374

3	4	9	8	6	7	2	1	5
7	6	2	3	1	5	9	8	4
5	8	1	2	9	4	7	3	6
8	3	7	5	2	6	1	4	9
9	2	5	4	8	1	6	7	3
6	1	4	7	3	9	5	2	8
1	7	3	9	5	8	4	6	2
4	5	8	6	7	2	3	9	1
2	9	6	1	4	3	8	5	7

Solution # 375

4	5	9	1	3	8	2	6	7
3	6	1	2	7	9	5	4	8
8	2	7	5	4	6	9	1	3
9	1	8	7	5	3	4	2	6
5	4	3	9	6	2	7	8	1
2	7	6	4	8	1	3	9	5
7	9	5	6	1	4	8	3	2
1	3	4	8	2	5	6	7	9
6	8	2	3	9	7	1	5	4

Solution # 376

6	9	4	1	3	8	2	7	5
7	5	1	2	4	9	3	6	8
8	2	3	5	7	6	4	9	1
2	1	9	4	6	5	7	8	3
5	8	7	3	1	2	9	4	6
4	3	6	9	8	7	1	5	2
3	7	5	8	2	4	6	1	9
9	4	2	6	5	1	8	3	7
1	6	8	7	9	3	5	2	4

Solution # 377

9	3	4	1	2	6	8	5	7
7	5	6	8	4	9	1	2	3
8	1	2	3	5	7	9	4	6
5	8	7	2	9	3	6	1	4
1	4	9	7	6	5	3	8	2
2	6	3	4	1	8	7	9	5
3	9	8	5	7	2	4	6	1
4	7	5	6	8	1	2	3	9
6	2	1	9	3	4	5	7	8

Solution # 378

5	2	9	7	1	3	4	8	6
3	4	6	2	9	8	7	5	1
8	7	1	6	4	5	9	3	2
4	3	8	9	5	6	1	2	7
7	9	5	1	3	2	8	6	4
1	6	2	8	7	4	5	9	3
6	5	3	4	8	7	2	1	9
9	8	7	3	2	1	6	4	5
2	1	4	5	6	9	3	7	8

Solution # 379

2	4	7	9	5	1	8	3	6
1	9	6	2	3	8	4	7	5
3	5	8	6	4	7	2	9	1
7	2	9	5	8	3	6	1	4
8	3	4	7	1	6	5	2	9
5	6	1	4	2	9	3	8	7
9	7	2	3	6	5	1	4	8
6	8	3	1	9	4	7	5	2
4	1	5	8	7	2	9	6	3

Solution # 380

8	6	7	5	2	3	4	1	9
4	2	3	1	9	8	6	7	5
5	9	1	4	6	7	2	3	8
7	8	6	3	5	4	9	2	1
1	5	4	2	7	9	3	8	6
9	3	2	8	1	6	7	5	4
3	7	8	6	4	1	5	9	2
6	1	5	9	3	2	8	4	7
2	4	9	7	8	5	1	6	3

Solution # 381

1	8	7	2	6	5	4	3	9
9	2	4	3	8	1	5	6	7
3	5	6	4	7	9	8	2	1
2	9	3	1	4	6	7	5	8
4	7	8	5	9	2	6	1	3
6	1	5	7	3	8	2	9	4
5	3	1	8	2	7	9	4	6
7	4	9	6	5	3	1	8	2
8	6	2	9	1	4	3	7	5

Solution # 382

4	6	8	3	2	7	9	1	5
5	3	7	9	4	1	8	2	6
2	1	9	5	8	6	4	3	7
8	2	3	6	7	5	1	9	4
6	7	4	1	9	2	3	5	8
9	5	1	4	3	8	6	7	2
1	4	6	2	5	9	7	8	3
3	8	5	7	1	4	2	6	9
7	9	2	8	6	3	5	4	1

Solution # 383

3	4	8	5	7	6	1	9	2
5	9	2	4	1	3	6	8	7
6	7	1	2	8	9	5	3	4
1	8	6	7	3	4	2	5	9
4	3	5	1	9	2	7	6	8
7	2	9	8	6	5	3	4	1
2	1	4	6	5	8	9	7	3
9	5	7	3	4	1	8	2	6
8	6	3	9	2	7	4	1	5

Solution # 384

3	6	1	7	4	8	5	9	2
7	8	9	5	2	3	4	6	1
4	5	2	6	1	9	8	3	7
9	4	5	3	7	6	2	1	8
2	3	7	9	8	1	6	5	4
8	1	6	4	5	2	3	7	9
6	7	4	2	9	5	1	8	3
1	2	3	8	6	7	9	4	5
5	9	8	1	3	4	7	2	6

Solution # 385

4	7	8	2	9	1	3	5	6
3	9	2	5	6	8	4	1	7
1	6	5	4	7	3	9	8	2
9	5	3	7	8	2	1	6	4
6	8	4	1	5	9	7	2	3
7	2	1	6	3	4	8	9	5
5	4	9	8	2	7	6	3	1
2	3	7	9	1	6	5	4	8
8	1	6	3	4	5	2	7	9

Solution # 386

4	9	7	6	5	1	8	2	3
5	3	1	8	2	4	6	9	7
2	6	8	9	3	7	5	4	1
6	2	4	3	1	8	7	5	9
3	7	5	4	9	6	2	1	8
8	1	9	5	7	2	3	6	4
9	4	3	7	6	5	1	8	2
1	8	6	2	4	3	9	7	5
7	5	2	1	8	9	4	3	6

Solution # 387

5	7	6	8	9	1	2	4	3
4	3	9	7	5	2	8	1	6
1	8	2	4	6	3	7	5	9
3	5	4	6	2	8	1	9	7
2	1	8	9	7	4	3	6	5
6	9	7	3	1	5	4	8	2
9	6	3	1	4	7	5	2	8
8	2	1	5	3	9	6	7	4
7	4	5	2	8	6	9	3	1

Solution # 388

7	6	4	8	5	3	2	9	1
1	3	8	6	2	9	7	5	4
9	5	2	7	1	4	6	8	3
6	4	9	3	8	5	1	2	7
2	8	3	9	7	1	5	4	6
5	1	7	2	4	6	9	3	8
8	9	5	4	6	7	3	1	2
3	2	6	1	9	8	4	7	5
4	7	1	5	3	2	8	6	9

Solution # 389

1	3	5	7	8	4	6	9	2
7	4	8	2	9	6	3	1	5
6	9	2	3	1	5	8	4	7
8	6	9	4	2	3	5	7	1
2	1	7	6	5	9	4	8	3
4	5	3	8	7	1	9	2	6
5	7	4	9	6	2	1	3	8
3	8	6	1	4	7	2	5	9
9	2	1	5	3	8	7	6	4

Solution # 390

3	9	2	5	4	1	7	6	8
8	5	6	7	9	2	1	3	4
4	1	7	6	3	8	2	9	5
6	4	5	9	1	7	3	8	2
9	8	3	2	5	4	6	7	1
7	2	1	8	6	3	4	5	9
5	7	9	4	2	6	8	1	3
1	6	4	3	8	5	9	2	7
2	3	8	1	7	9	5	4	6

Solution # 391

8	1	3	6	9	2	4	5	7
9	2	6	4	5	7	3	8	1
4	5	7	1	3	8	6	9	2
5	3	2	8	6	4	7	1	9
6	9	4	5	7	1	8	2	3
7	8	1	3	2	9	5	6	4
1	6	5	9	4	3	2	7	8
3	7	9	2	8	6	1	4	5
2	4	8	7	1	5	9	3	6

Solution # 392

8	4	6	5	3	1	2	7	9
3	2	7	4	6	9	8	1	5
9	5	1	7	2	8	6	3	4
6	9	3	8	4	7	5	2	1
4	7	5	6	1	2	3	9	8
1	8	2	9	5	3	7	4	6
5	6	9	2	7	4	1	8	3
2	3	8	1	9	5	4	6	7
7	1	4	3	8	6	9	5	2

Solution # 393

5	7	8	1	3	9	4	6	2
4	9	3	6	2	7	8	5	1
6	1	2	8	4	5	9	7	3
7	5	6	2	8	4	1	3	9
9	8	1	5	7	3	2	4	6
2	3	4	9	6	1	5	8	7
1	4	5	3	9	6	7	2	8
8	6	9	7	5	2	3	1	4
3	2	7	4	1	8	6	9	5

Solution # 394

4	1	2	3	5	7	8	9	6
7	8	3	9	6	1	2	5	4
6	5	9	2	4	8	7	1	3
9	4	5	8	7	3	6	2	1
2	7	8	6	1	4	5	3	9
3	6	1	5	2	9	4	7	8
1	2	4	7	9	6	3	8	5
8	9	7	4	3	5	1	6	2
5	3	6	1	8	2	9	4	7

Solution # 395

4	3	5	7	2	1	6	9	8
1	9	8	5	4	6	7	2	3
2	7	6	3	8	9	5	4	1
5	2	1	9	6	7	3	8	4
3	8	9	4	5	2	1	7	6
7	6	4	1	3	8	9	5	2
8	1	2	6	9	5	4	3	7
9	4	7	2	1	3	8	6	5
6	5	3	8	7	4	2	1	9

Solution # 396

1	2	4	8	6	9	5	3	7
5	7	8	2	1	3	4	9	6
3	6	9	4	5	7	2	8	1
2	4	1	3	9	8	7	6	5
9	5	6	1	7	2	3	4	8
7	8	3	5	4	6	9	1	2
8	1	5	7	3	4	6	2	9
6	3	2	9	8	5	1	7	4
4	9	7	6	2	1	8	5	3

Solution # 397

5	4	2	7	8	6	9	1	3
6	3	1	2	9	4	5	7	8
8	9	7	5	3	1	4	6	2
1	5	9	3	2	7	8	4	6
4	8	6	1	5	9	3	2	7
7	2	3	6	4	8	1	5	9
3	7	4	9	6	5	2	8	1
9	1	8	4	7	2	6	3	5
2	6	5	8	1	3	7	9	4

Solution # 398

4	3	7	5	1	6	9	2	8
5	8	1	9	2	4	7	6	3
2	6	9	8	7	3	4	5	1
9	1	6	3	8	7	5	4	2
7	5	8	2	4	1	3	9	6
3	2	4	6	5	9	1	8	7
1	9	5	7	6	8	2	3	4
6	4	3	1	9	2	8	7	5
8	7	2	4	3	5	6	1	9

Solution # 399

4	7	9	5	1	3	6	8	2
2	5	8	7	9	6	1	3	4
1	6	3	8	2	4	9	5	7
8	4	1	2	3	9	7	6	5
3	9	6	4	7	5	2	1	8
5	2	7	6	8	1	3	4	9
9	3	5	1	4	7	8	2	6
6	1	2	9	5	8	4	7	3
7	8	4	3	6	2	5	9	1

Solution # 400

3	8	4	1	9	7	6	2	5
2	1	6	5	8	4	7	9	3
5	9	7	6	3	2	1	8	4
1	7	3	4	2	6	9	5	8
8	6	5	3	7	9	2	4	1
9	4	2	8	1	5	3	7	6
4	5	9	7	6	1	8	3	2
7	3	1	2	4	8	5	6	9
6	2	8	9	5	3	4	1	7

Solution # 401

8	7	3	5	4	9	6	1	2
5	1	4	3	2	6	7	8	9
9	6	2	8	1	7	4	3	5
3	8	6	1	5	4	2	9	7
1	4	5	9	7	2	8	6	3
7	2	9	6	8	3	1	5	4
2	3	1	7	6	5	9	4	8
4	5	8	2	9	1	3	7	6
6	9	7	4	3	8	5	2	1

Solution # 402

8	9	6	7	5	2	1	4	3
4	2	1	3	8	6	9	5	7
5	3	7	1	4	9	8	2	6
9	7	8	4	2	3	5	6	1
3	4	2	5	6	1	7	9	8
1	6	5	8	9	7	4	3	2
7	5	4	6	3	8	2	1	9
2	8	3	9	1	5	6	7	4
6	1	9	2	7	4	3	8	5

Solution # 403

3	1	8	4	6	9	5	7	2
4	2	7	5	3	1	8	6	9
5	9	6	2	8	7	3	1	4
9	6	4	1	5	3	7	2	8
2	7	5	8	9	4	6	3	1
1	8	3	7	2	6	4	9	5
7	4	9	6	1	8	2	5	3
6	3	2	9	4	5	1	8	7
8	5	1	3	7	2	9	4	6

Solution # 404

1	5	6	7	3	4	8	2	9
9	7	2	1	8	5	3	4	6
8	4	3	2	9	6	7	5	1
3	8	4	6	2	9	1	7	5
5	1	7	3	4	8	9	6	2
2	6	9	5	1	7	4	8	3
4	2	5	9	7	3	6	1	8
7	9	1	8	6	2	5	3	4
6	3	8	4	5	1	2	9	7

Solution # 405

3	9	2	5	8	6	1	4	7
5	8	1	4	3	7	9	6	2
6	4	7	1	9	2	5	8	3
4	6	8	3	2	9	7	1	5
1	2	3	7	6	5	4	9	8
9	7	5	8	1	4	3	2	6
2	3	4	9	5	8	6	7	1
8	5	9	6	7	1	2	3	4
7	1	6	2	4	3	8	5	9

Solution # 406

4	3	9	6	2	8	7	1	5
7	5	8	1	4	3	2	6	9
1	6	2	9	7	5	4	8	3
8	9	5	4	6	7	3	2	1
2	4	6	5	3	1	8	9	7
3	1	7	2	8	9	5	4	6
5	8	4	7	9	6	1	3	2
6	7	3	8	1	2	9	5	4
9	2	1	3	5	4	6	7	8

Solution # 407

2	8	7	6	4	9	3	1	5
3	6	9	8	1	5	7	2	4
5	4	1	7	2	3	6	9	8
7	3	5	2	9	4	1	8	6
8	1	4	3	5	6	9	7	2
6	9	2	1	7	8	4	5	3
1	7	3	5	6	2	8	4	9
4	5	8	9	3	7	2	6	1
9	2	6	4	8	1	5	3	7

Solution # 408

4	2	8	9	6	1	7	3	5
5	9	7	3	2	4	8	6	1
6	3	1	7	8	5	2	9	4
9	6	5	1	3	2	4	8	7
2	7	3	8	4	9	1	5	6
8	1	4	5	7	6	3	2	9
1	4	2	6	5	8	9	7	3
3	8	6	4	9	7	5	1	2
7	5	9	2	1	3	6	4	8

Solution # 409

4	2	5	7	3	9	1	8	6
8	3	7	4	6	1	5	9	2
9	6	1	5	2	8	7	4	3
5	9	6	8	4	2	3	1	7
2	4	3	1	7	6	8	5	9
1	7	8	9	5	3	2	6	4
6	5	4	3	1	7	9	2	8
7	8	2	6	9	5	4	3	1
3	1	9	2	8	4	6	7	5

Solution # 410

8	9	5	6	7	1	4	3	2
4	7	1	2	3	9	8	6	5
6	2	3	5	4	8	9	7	1
3	5	4	8	6	2	7	1	9
9	1	8	3	5	7	6	2	4
7	6	2	1	9	4	5	8	3
2	3	7	9	8	5	1	4	6
5	8	6	4	1	3	2	9	7
1	4	9	7	2	6	3	5	8

Solution # 411

5	2	8	7	3	9	6	1	4
7	4	3	1	2	6	9	5	8
9	6	1	4	5	8	7	2	3
3	5	9	6	7	2	4	8	1
2	8	6	5	4	1	3	9	7
4	1	7	8	9	3	2	6	5
1	9	2	3	8	7	5	4	6
6	7	5	9	1	4	8	3	2
8	3	4	2	6	5	1	7	9

Solution # 412

4	2	8	7	6	9	5	1	3
5	1	7	4	3	2	6	9	8
9	3	6	5	8	1	7	2	4
3	4	1	6	9	7	2	8	5
8	6	5	2	1	4	9	3	7
7	9	2	8	5	3	4	6	1
1	7	4	3	2	6	8	5	9
6	5	3	9	4	8	1	7	2
2	8	9	1	7	5	3	4	6

Solution # 413

5	6	8	3	2	9	7	1	4
7	9	3	4	1	6	8	2	5
4	2	1	8	5	7	6	3	9
3	8	7	5	6	1	9	4	2
2	4	5	9	8	3	1	6	7
6	1	9	7	4	2	3	5	8
1	5	6	2	9	8	4	7	3
8	3	4	6	7	5	2	9	1
9	7	2	1	3	4	5	8	6

Solution # 414

5	9	6	3	4	1	2	8	7
1	7	2	5	8	6	4	9	3
8	4	3	9	7	2	6	5	1
4	8	5	1	6	3	7	2	9
7	2	1	8	9	4	3	6	5
3	6	9	7	2	5	8	1	4
2	3	8	4	5	9	1	7	6
6	5	4	2	1	7	9	3	8
9	1	7	6	3	8	5	4	2

Solution # 415

3	8	2	1	6	7	4	5	9
5	7	1	9	4	3	8	6	2
9	4	6	2	8	5	7	3	1
2	9	4	3	1	6	5	8	7
6	5	7	4	2	8	1	9	3
1	3	8	5	7	9	6	2	4
4	6	9	8	3	1	2	7	5
8	2	5	7	9	4	3	1	6
7	1	3	6	5	2	9	4	8

Solution # 416

7	3	1	9	2	6	5	8	4
8	9	6	1	4	5	2	3	7
4	2	5	3	8	7	1	6	9
1	6	7	2	9	3	8	4	5
2	5	8	6	1	4	9	7	3
3	4	9	7	5	8	6	2	1
5	7	2	4	6	1	3	9	8
9	1	4	8	3	2	7	5	6
6	8	3	5	7	9	4	1	2

Solution # 417

4	3	8	5	2	9	1	7	6
5	9	6	3	1	7	8	2	4
7	1	2	4	8	6	5	3	9
8	2	5	9	7	3	6	4	1
1	6	9	2	5	4	7	8	3
3	4	7	8	6	1	9	5	2
6	5	4	7	9	2	3	1	8
2	8	1	6	3	5	4	9	7
9	7	3	1	4	8	2	6	5

Solution # 418

6	9	3	2	5	8	7	4	1
5	8	1	7	3	4	2	9	6
7	2	4	1	9	6	8	3	5
1	6	9	4	7	2	3	5	8
8	4	5	3	1	9	6	2	7
2	3	7	6	8	5	4	1	9
3	1	6	9	4	7	5	8	2
4	7	8	5	2	1	9	6	3
9	5	2	8	6	3	1	7	4

Solution # 419

7	9	6	3	5	8	1	2	4
3	1	8	4	9	2	6	7	5
4	2	5	7	1	6	9	3	8
5	7	9	2	6	3	8	4	1
6	8	2	1	4	5	3	9	7
1	3	4	9	8	7	2	5	6
8	6	3	5	2	4	7	1	9
9	5	7	6	3	1	4	8	2
2	4	1	8	7	9	5	6	3

Solution # 420

8	4	5	9	7	1	2	6	3
2	1	6	8	3	5	7	9	4
3	9	7	4	2	6	1	5	8
9	8	2	1	4	7	6	3	5
5	7	4	3	6	9	8	2	1
1	6	3	2	5	8	4	7	9
4	2	9	7	8	3	5	1	6
6	3	8	5	1	2	9	4	7
7	5	1	6	9	4	3	8	2

Solution # 421

5	9	3	2	4	1	6	7	8
1	7	4	8	3	6	9	5	2
8	6	2	7	5	9	3	4	1
6	3	1	4	9	8	7	2	5
7	2	5	6	1	3	8	9	4
9	4	8	5	2	7	1	3	6
3	5	7	1	6	2	4	8	9
2	8	6	9	7	4	5	1	3
4	1	9	3	8	5	2	6	7

Solution # 422

8	1	2	3	7	6	9	4	5
9	7	3	4	5	2	6	1	8
4	6	5	9	1	8	7	2	3
2	9	4	8	6	7	3	5	1
1	5	7	2	9	3	8	6	4
3	8	6	1	4	5	2	9	7
6	4	1	7	8	9	5	3	2
7	3	9	5	2	1	4	8	6
5	2	8	6	3	4	1	7	9

Solution # 423

2	7	8	1	9	3	5	6	4
4	6	1	2	5	7	9	8	3
5	3	9	8	4	6	2	7	1
7	2	4	5	8	1	6	3	9
6	9	5	7	3	2	4	1	8
1	8	3	4	6	9	7	5	2
3	4	6	9	1	5	8	2	7
8	1	2	6	7	4	3	9	5
9	5	7	3	2	8	1	4	6

Solution # 424

5	7	6	4	9	1	3	2	8
9	4	3	8	2	7	1	6	5
1	2	8	3	5	6	7	9	4
2	5	7	1	4	3	6	8	9
8	9	4	7	6	5	2	1	3
3	6	1	2	8	9	5	4	7
7	3	2	9	1	4	8	5	6
6	8	9	5	7	2	4	3	1
4	1	5	6	3	8	9	7	2

Solution # 425

3	5	4	6	9	7	8	1	2
8	2	7	1	4	5	9	6	3
6	9	1	2	8	3	4	5	7
7	3	6	5	1	8	2	9	4
4	1	2	9	3	6	7	8	5
5	8	9	4	7	2	6	3	1
9	6	3	7	2	1	5	4	8
2	4	8	3	5	9	1	7	6
1	7	5	8	6	4	3	2	9

Solution # 426

4	1	9	2	7	5	6	8	3
6	7	2	3	8	4	5	1	9
3	5	8	9	1	6	4	2	7
7	4	1	8	2	3	9	6	5
2	3	6	1	5	9	8	7	4
9	8	5	6	4	7	2	3	1
1	2	4	5	3	8	7	9	6
5	9	3	7	6	2	1	4	8
8	6	7	4	9	1	3	5	2

Solution # 427

4	6	2	3	1	8	5	9	7
7	1	5	9	4	6	8	2	3
9	3	8	5	7	2	4	1	6
8	4	1	7	9	5	3	6	2
6	5	9	2	3	1	7	8	4
2	7	3	6	8	4	9	5	1
3	8	6	4	2	9	1	7	5
5	9	4	1	6	7	2	3	8
1	2	7	8	5	3	6	4	9

Solution # 428

8	2	6	3	5	4	9	7	1
1	4	5	9	6	7	2	8	3
7	3	9	8	1	2	5	6	4
9	1	4	2	8	6	3	5	7
3	5	8	7	9	1	4	2	6
2	6	7	5	4	3	8	1	9
5	7	1	4	3	8	6	9	2
6	9	3	1	2	5	7	4	8
4	8	2	6	7	9	1	3	5

Solution # 429

3	1	9	4	7	6	8	2	5
8	4	7	5	1	2	6	9	3
6	2	5	9	8	3	4	1	7
7	6	8	2	5	1	3	4	9
4	5	3	7	9	8	2	6	1
1	9	2	6	3	4	5	7	8
5	3	4	1	6	9	7	8	2
2	7	1	8	4	5	9	3	6
9	8	6	3	2	7	1	5	4

Solution # 430

7	2	8	5	3	1	4	6	9
3	9	4	2	7	6	8	1	5
6	1	5	8	9	4	7	2	3
8	6	9	7	5	2	1	3	4
1	5	7	9	4	3	2	8	6
2	4	3	1	6	8	9	5	7
4	8	6	3	2	9	5	7	1
9	7	2	6	1	5	3	4	8
5	3	1	4	8	7	6	9	2

Solution # 431

2	8	1	3	6	9	7	4	5
5	4	6	2	1	7	9	3	8
3	7	9	8	5	4	1	2	6
7	5	8	9	2	6	3	1	4
6	3	2	4	7	1	5	8	9
9	1	4	5	3	8	2	6	7
1	6	5	7	8	2	4	9	3
4	2	7	6	9	3	8	5	1
8	9	3	1	4	5	6	7	2

Solution # 432

9	2	6	5	1	4	7	3	8
7	5	3	9	8	6	2	1	4
8	1	4	3	7	2	5	6	9
2	4	7	8	3	5	1	9	6
1	3	9	6	4	7	8	2	5
5	6	8	1	2	9	4	7	3
4	8	2	7	6	3	9	5	1
3	7	5	4	9	1	6	8	2
6	9	1	2	5	8	3	4	7

Solution # 433

4	7	8	2	6	1	3	9	5
6	1	9	5	3	8	7	2	4
3	2	5	9	4	7	6	1	8
2	4	7	8	1	3	5	6	9
1	5	3	6	2	9	8	4	7
9	8	6	7	5	4	1	3	2
7	6	1	4	8	2	9	5	3
8	3	2	1	9	5	4	7	6
5	9	4	3	7	6	2	8	1

Solution # 434

7	1	6	4	8	9	5	3	2
4	5	9	3	2	1	7	8	6
2	3	8	7	5	6	9	4	1
9	7	1	8	3	5	2	6	4
6	2	3	9	7	4	8	1	5
5	8	4	1	6	2	3	9	7
1	4	5	2	9	3	6	7	8
3	6	7	5	4	8	1	2	9
8	9	2	6	1	7	4	5	3

Solution # 435

6	8	9	4	3	2	1	5	7
2	7	4	9	1	5	6	3	8
5	1	3	8	7	6	2	4	9
9	2	7	3	5	1	8	6	4
8	5	1	6	4	7	9	2	3
4	3	6	2	9	8	7	1	5
1	9	8	5	2	3	4	7	6
3	4	2	7	6	9	5	8	1
7	6	5	1	8	4	3	9	2

Solution # 436

5	7	4	1	2	3	6	8	9
8	2	9	5	4	6	7	3	1
3	1	6	8	9	7	5	2	4
9	5	7	2	3	4	1	6	8
6	3	2	9	1	8	4	7	5
1	4	8	6	7	5	2	9	3
4	6	3	7	8	1	9	5	2
2	8	5	4	6	9	3	1	7
7	9	1	3	5	2	8	4	6

Solution # 437

3	5	1	9	7	2	6	8	4
8	2	9	4	6	3	5	1	7
6	4	7	8	1	5	3	2	9
7	3	8	5	2	9	1	4	6
2	6	5	3	4	1	9	7	8
1	9	4	6	8	7	2	5	3
4	1	6	2	9	8	7	3	5
5	8	2	7	3	6	4	9	1
9	7	3	1	5	4	8	6	2

Solution # 438

7	9	8	4	5	3	2	1	6
1	3	5	6	8	2	4	7	9
2	6	4	1	7	9	5	8	3
9	1	6	2	4	7	3	5	8
8	2	3	5	6	1	9	4	7
5	4	7	9	3	8	6	2	1
4	8	2	7	9	6	1	3	5
6	7	1	3	2	5	8	9	4
3	5	9	8	1	4	7	6	2

Solution # 439

7	3	8	6	2	5	1	4	9
4	2	1	9	8	3	5	7	6
6	5	9	1	4	7	3	8	2
8	4	6	5	7	9	2	3	1
2	7	5	8	3	1	9	6	4
1	9	3	4	6	2	8	5	7
5	1	4	3	9	6	7	2	8
3	6	2	7	1	8	4	9	5
9	8	7	2	5	4	6	1	3

Solution # 440

6	8	3	5	1	9	4	7	2
7	9	2	8	6	4	1	3	5
4	5	1	3	2	7	9	8	6
3	7	5	4	9	2	8	6	1
2	4	8	1	5	6	3	9	7
9	1	6	7	3	8	2	5	4
1	3	7	9	4	5	6	2	8
8	2	4	6	7	3	5	1	9
5	6	9	2	8	1	7	4	3

Solution # 441

3	6	7	1	9	4	2	8	5
4	2	8	3	5	6	7	1	9
5	9	1	7	2	8	3	6	4
1	8	5	6	7	3	4	9	2
6	7	9	4	1	2	5	3	8
2	3	4	5	8	9	6	7	1
8	5	2	9	6	7	1	4	3
9	4	6	2	3	1	8	5	7
7	1	3	8	4	5	9	2	6

Solution # 442

7	1	9	2	6	8	5	3	4
3	8	4	5	7	1	6	9	2
2	5	6	9	4	3	8	7	1
4	9	3	7	8	2	1	6	5
1	2	5	6	3	9	7	4	8
6	7	8	4	1	5	3	2	9
5	6	7	1	9	4	2	8	3
8	4	2	3	5	7	9	1	6
9	3	1	8	2	6	4	5	7

Solution # 443

3	8	5	2	4	6	9	7	1
6	9	4	3	7	1	2	5	8
7	2	1	5	9	8	6	4	3
8	1	6	9	5	7	3	2	4
9	5	3	8	2	4	1	6	7
2	4	7	6	1	3	5	8	9
4	7	2	1	6	9	8	3	5
5	3	9	4	8	2	7	1	6
1	6	8	7	3	5	4	9	2

Solution # 444

4	5	8	1	9	2	3	7	6
1	6	2	3	7	4	5	9	8
7	3	9	5	8	6	4	1	2
6	1	5	4	2	9	8	3	7
2	9	7	8	1	3	6	4	5
8	4	3	7	6	5	1	2	9
5	7	4	2	3	8	9	6	1
9	8	1	6	4	7	2	5	3
3	2	6	9	5	1	7	8	4

Solution # 445

7	2	4	6	5	3	9	1	8
6	5	9	8	1	4	7	2	3
1	3	8	2	9	7	5	4	6
2	6	7	5	3	8	1	9	4
9	4	5	1	7	6	8	3	2
8	1	3	9	4	2	6	5	7
3	9	6	4	8	5	2	7	1
5	7	2	3	6	1	4	8	9
4	8	1	7	2	9	3	6	5

Solution # 446

3	5	8	4	6	1	7	2	9
9	4	7	3	2	5	1	8	6
6	1	2	9	8	7	4	5	3
8	2	6	5	7	3	9	1	4
5	7	9	2	1	4	3	6	8
1	3	4	6	9	8	5	7	2
4	8	1	7	3	6	2	9	5
2	6	5	1	4	9	8	3	7
7	9	3	8	5	2	6	4	1

Solution # 447

6	5	7	9	3	1	4	2	8
8	2	9	5	4	6	3	7	1
1	3	4	8	7	2	5	6	9
4	8	1	7	5	3	6	9	2
5	7	6	4	2	9	1	8	3
2	9	3	6	1	8	7	4	5
9	6	5	3	8	4	2	1	7
3	4	2	1	9	7	8	5	6
7	1	8	2	6	5	9	3	4

Solution # 448

7	8	9	2	4	1	5	6	3
6	1	4	9	3	5	8	7	2
2	5	3	7	6	8	4	9	1
3	2	6	8	5	7	9	1	4
8	9	5	4	1	3	6	2	7
4	7	1	6	9	2	3	5	8
9	3	8	1	2	6	7	4	5
5	4	2	3	7	9	1	8	6
1	6	7	5	8	4	2	3	9

Solution # 449

2	7	8	9	3	5	6	4	1
5	1	4	8	6	7	3	2	9
3	9	6	2	1	4	5	7	8
8	2	5	3	7	1	9	6	4
4	3	9	5	2	6	8	1	7
1	6	7	4	8	9	2	3	5
7	8	3	1	9	2	4	5	6
9	4	1	6	5	3	7	8	2
6	5	2	7	4	8	1	9	3

Solution # 450

3	8	7	4	1	9	5	6	2
5	4	1	3	6	2	9	8	7
6	9	2	7	8	5	4	1	3
4	7	3	9	2	6	8	5	1
8	5	9	1	4	3	7	2	6
1	2	6	5	7	8	3	9	4
7	3	8	6	9	1	2	4	5
9	1	4	2	5	7	6	3	8
2	6	5	8	3	4	1	7	9

Solution # 451

3	2	1	6	4	9	8	7	5
5	6	4	3	8	7	9	1	2
7	8	9	1	5	2	6	4	3
8	5	6	2	3	1	7	9	4
4	3	7	5	9	6	1	2	8
9	1	2	8	7	4	5	3	6
1	9	8	4	2	5	3	6	7
6	4	5	7	1	3	2	8	9
2	7	3	9	6	8	4	5	1

Solution # 452

3	9	4	2	6	7	1	5	8
2	5	7	1	4	8	6	3	9
8	1	6	5	3	9	4	7	2
5	3	2	6	9	1	8	4	7
7	6	8	3	5	4	2	9	1
9	4	1	7	8	2	5	6	3
4	7	3	8	2	6	9	1	5
1	2	9	4	7	5	3	8	6
6	8	5	9	1	3	7	2	4

Solution # 453

2	7	6	9	5	1	8	3	4
4	8	9	3	7	2	1	6	5
3	5	1	4	8	6	2	7	9
1	9	3	8	4	7	6	5	2
8	6	5	1	2	9	3	4	7
7	4	2	5	6	3	9	8	1
6	2	8	7	9	4	5	1	3
5	3	4	2	1	8	7	9	6
9	1	7	6	3	5	4	2	8

Solution # 454

4	9	2	7	6	3	1	5	8
1	6	8	5	2	9	7	3	4
5	7	3	4	1	8	6	9	2
8	2	5	3	7	4	9	6	1
7	4	1	2	9	6	5	8	3
9	3	6	8	5	1	2	4	7
6	8	9	1	3	2	4	7	5
3	1	7	6	4	5	8	2	9
2	5	4	9	8	7	3	1	6

Solution # 455

3	1	9	8	5	2	4	6	7
7	5	6	3	4	9	2	8	1
4	8	2	6	7	1	9	3	5
8	3	5	9	1	7	6	2	4
6	9	7	2	8	4	5	1	3
1	2	4	5	6	3	7	9	8
2	4	1	7	9	8	3	5	6
5	7	3	1	2	6	8	4	9
9	6	8	4	3	5	1	7	2

Solution # 456

9	4	2	3	8	1	5	6	7
8	1	5	2	6	7	4	3	9
7	3	6	5	4	9	8	1	2
6	9	8	1	5	3	2	7	4
5	7	1	4	2	6	9	8	3
4	2	3	7	9	8	6	5	1
3	5	9	8	7	4	1	2	6
2	6	7	9	1	5	3	4	8
1	8	4	6	3	2	7	9	5

Solution # 457

6	3	9	2	4	7	5	1	8
1	5	4	9	3	8	2	6	7
8	2	7	6	1	5	3	9	4
5	8	3	4	6	2	9	7	1
4	6	1	5	7	9	8	2	3
9	7	2	1	8	3	6	4	5
7	9	5	8	2	4	1	3	6
2	4	6	3	5	1	7	8	9
3	1	8	7	9	6	4	5	2

Solution # 458

8	2	9	1	6	4	3	5	7
5	6	3	9	8	7	2	1	4
1	7	4	5	2	3	8	6	9
3	4	6	2	1	8	7	9	5
7	5	2	3	4	9	6	8	1
9	8	1	7	5	6	4	2	3
6	3	7	8	9	1	5	4	2
2	1	8	4	7	5	9	3	6
4	9	5	6	3	2	1	7	8

Solution # 459

2	5	4	8	6	3	7	1	9
9	6	8	5	7	1	4	2	3
3	1	7	2	9	4	6	5	8
6	2	9	1	4	8	3	7	5
8	3	1	6	5	7	9	4	2
4	7	5	9	3	2	1	8	6
1	9	2	7	8	6	5	3	4
7	4	6	3	2	5	8	9	1
5	8	3	4	1	9	2	6	7

Solution # 460

3	6	9	1	4	7	5	2	8
7	5	1	2	3	8	6	9	4
2	4	8	5	6	9	3	7	1
4	8	3	6	2	1	9	5	7
5	2	7	3	9	4	8	1	6
9	1	6	7	8	5	2	4	3
1	9	2	8	7	6	4	3	5
8	7	4	9	5	3	1	6	2
6	3	5	4	1	2	7	8	9

Solution # 461

6	8	1	7	5	9	2	3	4
7	3	9	4	2	1	8	6	5
4	5	2	3	6	8	7	9	1
8	1	4	2	7	3	9	5	6
2	6	7	9	4	5	1	8	3
5	9	3	8	1	6	4	7	2
9	2	8	5	3	4	6	1	7
1	4	5	6	8	7	3	2	9
3	7	6	1	9	2	5	4	8

Solution # 462

2	8	4	7	3	9	5	1	6
1	6	3	5	4	8	9	7	2
9	5	7	6	2	1	3	4	8
8	1	5	9	6	2	4	3	7
3	7	2	4	1	5	8	6	9
6	4	9	3	8	7	2	5	1
4	3	1	8	9	6	7	2	5
5	2	8	1	7	3	6	9	4
7	9	6	2	5	4	1	8	3

Solution # 463

6	7	9	8	3	2	1	5	4
3	5	8	4	7	1	9	6	2
4	2	1	9	6	5	8	3	7
7	1	6	5	8	9	2	4	3
5	8	2	6	4	3	7	9	1
9	3	4	1	2	7	5	8	6
8	6	7	2	9	4	3	1	5
1	4	3	7	5	8	6	2	9
2	9	5	3	1	6	4	7	8

Solution # 464

8	4	7	1	6	5	2	9	3
6	5	3	9	8	2	1	4	7
9	1	2	3	7	4	5	8	6
2	3	1	7	5	9	8	6	4
7	9	4	8	1	6	3	2	5
5	8	6	4	2	3	9	7	1
1	7	9	5	4	8	6	3	2
4	2	8	6	3	1	7	5	9
3	6	5	2	9	7	4	1	8

Solution # 465

4	9	3	2	5	6	1	7	8
7	5	2	8	9	1	3	4	6
8	6	1	3	4	7	9	2	5
3	4	8	6	7	9	5	1	2
1	7	9	5	2	8	6	3	4
5	2	6	4	1	3	8	9	7
2	1	7	9	6	5	4	8	3
6	8	4	1	3	2	7	5	9
9	3	5	7	8	4	2	6	1

Solution # 466

4	8	3	2	7	5	6	9	1
2	5	9	3	1	6	7	8	4
7	6	1	9	8	4	5	3	2
3	2	8	4	5	1	9	6	7
1	9	7	6	2	3	4	5	8
5	4	6	7	9	8	2	1	3
9	3	4	1	6	7	8	2	5
8	7	2	5	3	9	1	4	6
6	1	5	8	4	2	3	7	9

Solution # 467

3	7	4	6	2	8	9	1	5
2	9	8	4	1	5	3	7	6
5	1	6	7	9	3	4	8	2
9	2	1	3	8	4	5	6	7
8	5	7	9	6	2	1	3	4
4	6	3	1	5	7	2	9	8
1	4	5	8	3	6	7	2	9
7	8	9	2	4	1	6	5	3
6	3	2	5	7	9	8	4	1

Solution # 468

7	9	1	5	6	3	8	2	4
2	3	8	9	7	4	5	6	1
6	5	4	2	8	1	7	9	3
9	6	7	4	1	2	3	5	8
3	4	2	8	5	6	1	7	9
8	1	5	7	3	9	2	4	6
5	7	6	1	9	8	4	3	2
1	2	3	6	4	5	9	8	7
4	8	9	3	2	7	6	1	5

Solution # 469

6	3	8	7	5	9	1	4	2
1	5	2	4	6	3	7	8	9
4	7	9	1	8	2	3	5	6
7	6	5	8	9	4	2	3	1
2	8	3	5	7	1	6	9	4
9	1	4	2	3	6	5	7	8
3	2	7	6	4	8	9	1	5
5	4	6	9	1	7	8	2	3
8	9	1	3	2	5	4	6	7

Solution # 470

3	4	9	7	2	6	5	8	1
8	5	2	1	3	4	7	9	6
6	1	7	8	9	5	4	3	2
5	8	3	6	7	2	9	1	4
7	9	6	5	4	1	3	2	8
1	2	4	3	8	9	6	5	7
2	3	1	9	6	7	8	4	5
9	6	5	4	1	8	2	7	3
4	7	8	2	5	3	1	6	9

Solution # 471

4	1	6	5	9	2	3	7	8
7	9	3	8	4	1	2	6	5
2	8	5	3	6	7	9	4	1
6	7	1	9	3	5	4	8	2
8	4	9	2	1	6	5	3	7
3	5	2	7	8	4	1	9	6
9	3	7	1	2	8	6	5	4
5	2	4	6	7	9	8	1	3
1	6	8	4	5	3	7	2	9

Solution # 472

1	8	2	9	7	3	5	6	4
3	6	5	8	2	4	1	9	7
4	9	7	5	1	6	3	8	2
2	1	8	4	3	9	7	5	6
7	4	3	2	6	5	8	1	9
9	5	6	7	8	1	2	4	3
5	3	1	6	4	7	9	2	8
6	2	9	3	5	8	4	7	1
8	7	4	1	9	2	6	3	5

Solution # 473

1	5	6	3	2	9	4	8	7
7	4	3	6	8	1	2	9	5
8	2	9	5	7	4	3	1	6
4	9	5	8	6	3	1	7	2
6	3	1	2	9	7	8	5	4
2	8	7	4	1	5	6	3	9
3	7	4	1	5	2	9	6	8
5	6	2	9	3	8	7	4	1
9	1	8	7	4	6	5	2	3

Solution # 474

9	2	8	5	4	3	7	6	1
1	5	7	2	8	6	9	3	4
6	4	3	9	7	1	5	8	2
2	3	4	6	5	7	8	1	9
7	9	1	4	2	8	3	5	6
5	8	6	1	3	9	4	2	7
4	7	5	8	1	2	6	9	3
3	6	2	7	9	5	1	4	8
8	1	9	3	6	4	2	7	5

Solution # 475

7	8	6	5	9	1	4	2	3
2	9	5	6	4	3	7	8	1
4	1	3	7	2	8	5	9	6
6	3	2	1	5	7	8	4	9
1	5	8	4	3	9	6	7	2
9	4	7	8	6	2	1	3	5
3	7	1	2	8	5	9	6	4
5	6	9	3	7	4	2	1	8
8	2	4	9	1	6	3	5	7

Solution # 476

4	1	3	5	9	6	2	8	7
5	6	7	8	2	3	4	9	1
2	8	9	4	7	1	3	5	6
7	2	4	3	5	9	1	6	8
1	3	6	7	4	8	5	2	9
8	9	5	1	6	2	7	4	3
3	7	2	9	8	5	6	1	4
9	5	1	6	3	4	8	7	2
6	4	8	2	1	7	9	3	5

Solution # 477

3	6	2	9	7	1	8	4	5
9	1	7	4	8	5	3	6	2
5	4	8	2	6	3	9	7	1
1	7	5	8	2	9	4	3	6
8	2	3	6	1	4	5	9	7
6	9	4	3	5	7	1	2	8
2	3	6	5	4	8	7	1	9
4	8	1	7	9	6	2	5	3
7	5	9	1	3	2	6	8	4

Solution # 478

1	4	3	5	8	6	2	7	9
2	5	7	4	1	9	8	6	3
8	6	9	2	7	3	5	4	1
4	3	8	7	9	1	6	5	2
7	1	5	6	4	2	3	9	8
9	2	6	3	5	8	4	1	7
5	7	2	1	3	4	9	8	6
6	9	4	8	2	7	1	3	5
3	8	1	9	6	5	7	2	4

Solution # 479

9	4	1	5	2	7	3	6	8
8	5	2	4	6	3	9	1	7
3	7	6	1	9	8	4	5	2
6	1	4	3	8	5	7	2	9
7	8	3	9	1	2	6	4	5
5	2	9	7	4	6	8	3	1
1	9	8	2	3	4	5	7	6
4	6	7	8	5	1	2	9	3
2	3	5	6	7	9	1	8	4

Solution # 480

1	8	6	4	9	3	5	2	7
2	7	3	1	5	6	8	9	4
5	9	4	2	8	7	1	3	6
7	3	9	5	6	1	4	8	2
8	2	1	7	3	4	9	6	5
6	4	5	8	2	9	7	1	3
4	6	8	3	1	5	2	7	9
3	1	7	9	4	2	6	5	8
9	5	2	6	7	8	3	4	1

Solution # 481

5	1	2	4	3	7	8	6	9
7	4	9	5	6	8	1	2	3
6	8	3	1	2	9	7	5	4
9	3	7	2	8	1	5	4	6
2	5	8	6	4	3	9	7	1
4	6	1	7	9	5	3	8	2
8	7	4	3	1	6	2	9	5
1	2	5	9	7	4	6	3	8
3	9	6	8	5	2	4	1	7

Solution # 482

2	5	7	8	6	1	4	9	3
8	9	3	2	5	4	6	7	1
6	4	1	9	3	7	2	8	5
3	7	5	4	1	6	9	2	8
9	8	4	5	7	2	3	1	6
1	2	6	3	8	9	5	4	7
5	1	2	7	4	3	8	6	9
7	3	9	6	2	8	1	5	4
4	6	8	1	9	5	7	3	2

Solution # 483

8	1	5	9	7	6	3	4	2
4	6	2	5	1	3	9	8	7
9	7	3	4	2	8	6	1	5
7	5	6	1	9	2	8	3	4
2	4	9	8	3	7	5	6	1
3	8	1	6	4	5	7	2	9
1	3	4	7	6	9	2	5	8
5	2	7	3	8	1	4	9	6
6	9	8	2	5	4	1	7	3

Solution # 484

4	8	6	2	9	7	5	3	1
5	2	7	1	4	3	9	6	8
3	1	9	6	8	5	4	2	7
6	5	1	9	3	2	8	7	4
8	3	2	4	7	1	6	9	5
7	9	4	8	5	6	2	1	3
1	4	5	7	6	9	3	8	2
2	6	3	5	1	8	7	4	9
9	7	8	3	2	4	1	5	6

Solution # 485

9	6	8	4	3	2	5	7	1
7	4	2	1	5	8	9	3	6
5	3	1	9	7	6	8	4	2
8	5	7	6	9	4	2	1	3
2	9	3	7	1	5	6	8	4
6	1	4	8	2	3	7	9	5
3	7	6	5	8	1	4	2	9
1	8	5	2	4	9	3	6	7
4	2	9	3	6	7	1	5	8

Solution # 486

4	6	9	8	3	5	7	2	1
5	8	1	9	7	2	3	6	4
3	2	7	1	4	6	5	8	9
2	7	3	6	1	4	9	5	8
8	1	4	5	9	3	2	7	6
6	9	5	7	2	8	1	4	3
1	5	8	2	6	9	4	3	7
9	3	2	4	8	7	6	1	5
7	4	6	3	5	1	8	9	2

Solution # 487

4	7	8	1	5	2	6	9	3
5	9	1	7	3	6	4	2	8
3	2	6	9	8	4	5	7	1
1	5	9	8	6	3	7	4	2
2	8	7	4	1	5	9	3	6
6	4	3	2	9	7	8	1	5
8	6	4	3	7	1	2	5	9
7	3	5	6	2	9	1	8	4
9	1	2	5	4	8	3	6	7

Solution # 488

5	7	9	6	4	1	8	3	2
4	6	8	3	7	2	5	1	9
1	2	3	5	9	8	6	7	4
2	5	4	8	6	7	1	9	3
8	3	1	9	2	4	7	6	5
6	9	7	1	5	3	4	2	8
3	4	6	7	8	9	2	5	1
7	1	2	4	3	5	9	8	6
9	8	5	2	1	6	3	4	7

Solution # 489

5	8	2	9	4	1	3	7	6
9	1	6	7	3	5	2	4	8
7	3	4	6	8	2	1	5	9
6	4	9	3	1	8	5	2	7
1	2	8	5	9	7	4	6	3
3	5	7	2	6	4	8	9	1
2	6	3	8	5	9	7	1	4
8	7	1	4	2	6	9	3	5
4	9	5	1	7	3	6	8	2

Solution # 490

8	3	2	5	1	6	4	9	7
4	9	6	7	2	3	5	8	1
1	7	5	9	4	8	3	6	2
2	5	4	6	9	1	7	3	8
9	1	8	3	7	5	2	4	6
3	6	7	2	8	4	9	1	5
7	8	9	1	3	2	6	5	4
5	2	1	4	6	9	8	7	3
6	4	3	8	5	7	1	2	9

Solution # 491

5	4	3	8	2	9	7	6	1
8	1	2	3	6	7	5	9	4
7	6	9	5	1	4	2	3	8
6	7	8	2	4	3	9	1	5
9	2	1	6	8	5	3	4	7
4	3	5	7	9	1	8	2	6
2	5	6	1	3	8	4	7	9
1	8	4	9	7	2	6	5	3
3	9	7	4	5	6	1	8	2

Solution # 492

4	6	9	5	1	3	8	7	2
2	3	5	9	7	8	4	6	1
7	1	8	6	4	2	5	9	3
9	4	7	2	3	5	1	8	6
3	2	6	1	8	7	9	4	5
5	8	1	4	6	9	3	2	7
8	5	3	7	2	4	6	1	9
1	7	4	3	9	6	2	5	8
6	9	2	8	5	1	7	3	4

Solution # 493

7	4	6	5	3	8	9	1	2
3	9	1	4	6	2	5	7	8
5	8	2	7	9	1	6	3	4
2	5	7	3	1	6	8	4	9
8	3	4	9	2	5	7	6	1
6	1	9	8	4	7	3	2	5
1	7	5	2	8	3	4	9	6
9	6	8	1	7	4	2	5	3
4	2	3	6	5	9	1	8	7

Solution # 494

5	6	3	2	7	1	9	8	4
9	8	4	6	3	5	1	2	7
2	7	1	8	9	4	5	3	6
1	2	7	3	8	9	6	4	5
6	5	8	7	4	2	3	9	1
3	4	9	5	1	6	8	7	2
7	1	2	9	5	3	4	6	8
8	9	5	4	6	7	2	1	3
4	3	6	1	2	8	7	5	9

Solution # 495

7	8	3	5	6	4	1	9	2
5	4	2	7	1	9	6	3	8
6	9	1	3	2	8	4	5	7
4	3	6	8	9	2	7	1	5
8	5	7	4	3	1	9	2	6
2	1	9	6	5	7	3	8	4
3	2	5	9	7	6	8	4	1
1	6	4	2	8	3	5	7	9
9	7	8	1	4	5	2	6	3

Solution # 496

3	5	6	1	8	7	4	9	2
2	4	1	6	3	9	7	5	8
9	8	7	4	5	2	3	6	1
4	7	9	3	2	1	5	8	6
5	3	8	9	4	6	2	1	7
6	1	2	8	7	5	9	3	4
8	9	3	7	1	4	6	2	5
1	2	4	5	6	3	8	7	9
7	6	5	2	9	8	1	4	3

Solution # 497

1	3	5	4	8	6	7	2	9
9	2	4	3	5	7	8	6	1
8	6	7	2	1	9	5	4	3
4	8	6	9	2	5	3	1	7
5	7	1	6	3	4	9	8	2
3	9	2	8	7	1	6	5	4
7	4	9	1	6	8	2	3	5
6	5	3	7	4	2	1	9	8
2	1	8	5	9	3	4	7	6

Solution # 498

3	4	8	9	1	6	7	5	2
5	7	6	3	4	2	1	9	8
2	1	9	8	5	7	6	4	3
7	3	4	2	8	9	5	1	6
9	8	2	1	6	5	3	7	4
1	6	5	4	7	3	8	2	9
6	2	1	5	3	4	9	8	7
4	5	7	6	9	8	2	3	1
8	9	3	7	2	1	4	6	5

Solution # 499

9	8	5	4	1	6	3	7	2
4	2	6	7	5	3	9	1	8
7	1	3	2	8	9	4	5	6
3	4	9	6	2	5	1	8	7
1	6	2	8	4	7	5	9	3
8	5	7	3	9	1	6	2	4
5	7	8	1	3	4	2	6	9
6	3	1	9	7	2	8	4	5
2	9	4	5	6	8	7	3	1

Solution # 500

3	5	2	4	9	6	8	1	7
6	7	9	1	8	2	5	3	4
8	4	1	5	7	3	2	9	6
2	9	8	3	5	4	6	7	1
4	1	5	7	6	8	3	2	9
7	6	3	2	1	9	4	8	5
9	2	6	8	4	7	1	5	3
1	8	7	6	3	5	9	4	2
5	3	4	9	2	1	7	6	8

Solution # 501

5	2	1	9	4	3	6	7	8
4	9	6	7	8	1	3	2	5
3	7	8	2	5	6	9	4	1
6	4	5	8	3	7	2	1	9
9	3	7	6	1	2	5	8	4
1	8	2	5	9	4	7	6	3
8	6	9	1	7	5	4	3	2
7	5	4	3	2	8	1	9	6
2	1	3	4	6	9	8	5	7

Solution # 502

2	4	9	3	5	1	8	7	6
3	6	7	8	4	9	1	5	2
8	5	1	7	6	2	4	3	9
1	7	5	6	3	4	9	2	8
6	3	8	9	2	5	7	4	1
9	2	4	1	7	8	5	6	3
7	1	2	4	8	3	6	9	5
5	9	6	2	1	7	3	8	4
4	8	3	5	9	6	2	1	7

Solution # 503

9	5	4	6	2	8	7	3	1
1	3	6	5	9	7	4	8	2
7	8	2	1	4	3	9	5	6
2	6	7	4	8	1	3	9	5
3	4	5	9	7	2	6	1	8
8	9	1	3	6	5	2	4	7
6	1	8	7	3	9	5	2	4
4	2	9	8	5	6	1	7	3
5	7	3	2	1	4	8	6	9

Solution # 504

8	7	9	5	4	6	3	1	2
4	3	5	2	1	7	6	9	8
2	1	6	9	3	8	7	4	5
9	8	3	1	2	5	4	6	7
6	4	1	8	7	9	2	5	3
7	5	2	4	6	3	9	8	1
1	6	4	3	8	2	5	7	9
3	9	8	7	5	4	1	2	6
5	2	7	6	9	1	8	3	4

Solution # 505

7	5	1	2	6	9	8	4	3
8	2	3	1	5	4	9	6	7
6	4	9	7	3	8	1	5	2
1	6	4	3	2	5	7	8	9
9	7	2	4	8	6	3	1	5
3	8	5	9	1	7	6	2	4
4	3	6	8	7	2	5	9	1
2	1	8	5	9	3	4	7	6
5	9	7	6	4	1	2	3	8

Solution # 506

5	6	7	9	2	1	3	4	8
8	9	3	4	5	6	2	7	1
4	1	2	3	8	7	9	6	5
2	5	1	8	7	9	6	3	4
6	3	9	5	1	4	7	8	2
7	4	8	2	6	3	5	1	9
9	2	6	7	4	8	1	5	3
3	7	4	1	9	5	8	2	6
1	8	5	6	3	2	4	9	7

Solution # 507

9	7	5	8	1	2	6	4	3
4	6	2	9	5	3	8	1	7
8	1	3	7	4	6	5	9	2
1	2	6	5	7	8	4	3	9
5	3	4	1	2	9	7	6	8
7	9	8	3	6	4	1	2	5
2	8	1	6	3	5	9	7	4
3	5	7	4	9	1	2	8	6
6	4	9	2	8	7	3	5	1

Solution # 508

3	1	2	8	4	9	5	7	6
4	7	5	1	2	6	8	9	3
6	8	9	3	5	7	1	4	2
7	9	3	5	6	4	2	8	1
2	5	6	9	8	1	7	3	4
8	4	1	2	7	3	9	6	5
1	3	4	7	9	2	6	5	8
9	2	8	6	3	5	4	1	7
5	6	7	4	1	8	3	2	9

Solution # 509

5	6	1	3	8	2	4	7	9
9	7	3	4	5	6	1	8	2
2	4	8	7	1	9	5	6	3
8	9	2	6	7	1	3	4	5
7	5	6	2	3	4	9	1	8
1	3	4	8	9	5	7	2	6
6	1	9	5	4	8	2	3	7
3	2	5	1	6	7	8	9	4
4	8	7	9	2	3	6	5	1

Solution # 510

5	2	6	4	7	8	9	1	3
3	7	8	1	2	9	5	6	4
9	1	4	3	5	6	2	8	7
8	4	9	6	3	1	7	5	2
2	5	1	9	4	7	6	3	8
7	6	3	2	8	5	4	9	1
1	8	5	7	9	4	3	2	6
4	9	2	8	6	3	1	7	5
6	3	7	5	1	2	8	4	9

Solution # 511

1	8	9	5	4	6	2	3	7
7	4	5	8	3	2	9	1	6
2	3	6	1	7	9	4	8	5
3	5	1	4	8	7	6	9	2
4	6	2	9	1	5	3	7	8
8	9	7	2	6	3	5	4	1
9	7	8	6	5	4	1	2	3
6	1	4	3	2	8	7	5	9
5	2	3	7	9	1	8	6	4

Solution # 512

2	7	9	4	5	3	1	8	6
8	6	4	2	9	1	3	5	7
5	3	1	7	8	6	9	4	2
4	8	5	9	6	7	2	1	3
6	2	3	8	1	5	7	9	4
1	9	7	3	4	2	8	6	5
3	1	6	5	7	8	4	2	9
7	4	8	6	2	9	5	3	1
9	5	2	1	3	4	6	7	8

Solution # 513

4	5	3	1	6	9	2	7	8
9	2	7	8	5	4	3	6	1
1	8	6	3	2	7	9	5	4
8	6	9	5	4	3	1	2	7
7	4	2	6	9	1	5	8	3
3	1	5	2	7	8	6	4	9
2	7	8	9	3	6	4	1	5
5	9	1	4	8	2	7	3	6
6	3	4	7	1	5	8	9	2

Solution # 514

3	8	7	6	4	9	2	5	1
2	5	6	1	8	7	3	9	4
9	1	4	3	5	2	7	6	8
1	2	5	7	6	4	8	3	9
7	9	3	8	1	5	6	4	2
6	4	8	9	2	3	5	1	7
8	3	1	2	9	6	4	7	5
4	6	2	5	7	1	9	8	3
5	7	9	4	3	8	1	2	6

Solution # 515

1	3	5	9	6	8	4	7	2
4	9	2	1	7	5	8	3	6
7	8	6	3	2	4	1	9	5
8	5	9	4	3	6	2	1	7
6	7	4	5	1	2	3	8	9
3	2	1	8	9	7	6	5	4
9	4	7	2	8	3	5	6	1
2	1	3	6	5	9	7	4	8
5	6	8	7	4	1	9	2	3

Solution # 516

7	5	9	8	3	2	4	1	6
4	1	3	5	6	9	8	7	2
8	2	6	4	1	7	3	9	5
3	4	7	2	8	1	5	6	9
5	6	2	9	7	4	1	8	3
1	9	8	6	5	3	7	2	4
6	3	4	7	9	8	2	5	1
2	7	5	1	4	6	9	3	8
9	8	1	3	2	5	6	4	7

Solution # 517

7	5	3	8	6	9	4	2	1
4	8	1	2	3	5	7	6	9
2	6	9	4	1	7	3	8	5
9	2	7	3	5	4	6	1	8
3	4	6	1	9	8	5	7	2
8	1	5	6	7	2	9	4	3
5	3	8	7	2	6	1	9	4
6	9	4	5	8	1	2	3	7
1	7	2	9	4	3	8	5	6

Solution # 518

2	8	6	9	5	7	1	3	4
5	1	9	4	3	2	7	8	6
7	3	4	8	1	6	5	9	2
8	9	2	5	4	3	6	1	7
1	4	5	7	6	9	3	2	8
6	7	3	1	2	8	4	5	9
4	5	8	2	7	1	9	6	3
3	2	1	6	9	4	8	7	5
9	6	7	3	8	5	2	4	1

Solution # 519

2	1	3	9	6	7	4	5	8
6	7	5	4	3	8	9	1	2
8	4	9	5	2	1	3	6	7
9	5	2	1	8	3	7	4	6
7	8	4	2	5	6	1	9	3
3	6	1	7	9	4	2	8	5
5	3	7	8	1	9	6	2	4
4	9	8	6	7	2	5	3	1
1	2	6	3	4	5	8	7	9

Solution # 520

5	6	9	3	8	7	2	1	4
4	3	1	5	6	2	7	8	9
8	2	7	9	4	1	3	6	5
3	7	6	1	2	9	4	5	8
9	4	2	7	5	8	1	3	6
1	8	5	4	3	6	9	7	2
6	1	8	2	7	4	5	9	3
2	9	3	6	1	5	8	4	7
7	5	4	8	9	3	6	2	1

Solution # 521

3	6	1	8	4	9	5	2	7
4	7	9	1	2	5	3	8	6
2	5	8	6	7	3	4	9	1
6	4	3	2	8	1	7	5	9
1	2	7	5	9	6	8	4	3
9	8	5	4	3	7	6	1	2
8	9	4	3	6	2	1	7	5
7	1	6	9	5	8	2	3	4
5	3	2	7	1	4	9	6	8

Solution # 522

3	6	8	7	9	1	2	5	4
4	9	2	5	6	3	1	8	7
5	7	1	4	2	8	9	6	3
1	5	7	8	4	9	6	3	2
9	4	6	1	3	2	8	7	5
8	2	3	6	5	7	4	1	9
6	1	4	2	7	5	3	9	8
2	3	5	9	8	6	7	4	1
7	8	9	3	1	4	5	2	6

Solution # 523

6	8	9	2	4	5	7	1	3
2	3	7	9	6	1	8	5	4
5	4	1	8	3	7	6	9	2
4	1	3	6	5	2	9	7	8
8	7	5	1	9	4	3	2	6
9	6	2	3	7	8	5	4	1
3	5	6	4	1	9	2	8	7
7	2	4	5	8	6	1	3	9
1	9	8	7	2	3	4	6	5

Solution # 524

5	3	4	2	1	7	9	6	8
2	8	6	5	4	9	1	3	7
9	1	7	8	6	3	4	2	5
1	5	8	9	2	6	3	7	4
7	6	2	4	3	1	5	8	9
3	4	9	7	5	8	2	1	6
8	2	3	6	9	4	7	5	1
6	9	5	1	7	2	8	4	3
4	7	1	3	8	5	6	9	2

Solution # 525

8	7	2	1	3	6	9	4	5
9	5	4	7	2	8	3	6	1
1	3	6	4	5	9	7	2	8
7	6	1	2	9	4	5	8	3
3	9	5	6	8	7	2	1	4
2	4	8	3	1	5	6	7	9
4	8	9	5	6	2	1	3	7
5	2	3	8	7	1	4	9	6
6	1	7	9	4	3	8	5	2

Solution # 526

5	2	9	3	4	7	1	6	8
3	7	6	1	5	8	4	2	9
4	1	8	2	6	9	3	5	7
2	5	7	4	3	6	8	9	1
9	4	1	8	2	5	6	7	3
6	8	3	9	7	1	5	4	2
1	9	5	7	8	4	2	3	6
8	3	4	6	9	2	7	1	5
7	6	2	5	1	3	9	8	4

Solution # 527

8	5	2	9	4	3	6	1	7
4	6	1	2	7	8	3	9	5
9	7	3	5	1	6	8	4	2
2	4	9	8	5	1	7	6	3
7	1	5	3	6	9	2	8	4
6	3	8	4	2	7	9	5	1
1	2	6	7	9	4	5	3	8
3	9	7	1	8	5	4	2	6
5	8	4	6	3	2	1	7	9

Solution # 528

9	6	1	3	4	7	5	8	2
2	4	8	9	5	1	3	7	6
5	7	3	2	8	6	1	4	9
7	3	2	8	1	9	6	5	4
1	5	9	4	6	3	7	2	8
4	8	6	5	7	2	9	3	1
3	2	5	6	9	8	4	1	7
6	1	4	7	2	5	8	9	3
8	9	7	1	3	4	2	6	5

Solution # 529

2	4	5	7	8	1	6	9	3
1	6	8	5	3	9	2	7	4
3	7	9	2	6	4	5	8	1
4	5	1	3	9	6	8	2	7
9	3	7	8	5	2	1	4	6
8	2	6	1	4	7	9	3	5
6	8	4	9	7	5	3	1	2
7	9	2	6	1	3	4	5	8
5	1	3	4	2	8	7	6	9

Solution # 530

9	4	6	5	2	3	1	8	7
3	7	1	8	9	4	2	6	5
8	5	2	7	1	6	9	3	4
7	2	8	6	5	9	3	4	1
5	9	3	4	8	1	7	2	6
1	6	4	3	7	2	5	9	8
4	1	5	2	3	8	6	7	9
6	3	7	9	4	5	8	1	2
2	8	9	1	6	7	4	5	3

Solution # 531

7	4	1	3	9	8	6	5	2
2	6	9	7	4	5	8	3	1
3	8	5	2	6	1	9	7	4
8	5	2	4	3	7	1	6	9
4	3	6	1	2	9	7	8	5
9	1	7	8	5	6	2	4	3
5	7	8	9	1	3	4	2	6
6	9	4	5	7	2	3	1	8
1	2	3	6	8	4	5	9	7

Solution # 532

6	7	3	5	1	8	9	2	4
4	9	2	7	6	3	8	1	5
8	1	5	4	2	9	6	7	3
5	3	8	2	4	7	1	9	6
9	2	6	1	3	5	7	4	8
1	4	7	9	8	6	5	3	2
3	5	4	8	9	1	2	6	7
2	8	1	6	7	4	3	5	9
7	6	9	3	5	2	4	8	1

Solution # 533

4	9	8	6	5	2	7	3	1
6	7	5	1	9	3	8	4	2
1	3	2	7	8	4	5	6	9
3	2	9	5	4	1	6	8	7
5	1	6	8	7	9	3	2	4
8	4	7	3	2	6	9	1	5
7	8	4	2	3	5	1	9	6
9	5	1	4	6	8	2	7	3
2	6	3	9	1	7	4	5	8

Solution # 534

7	5	6	8	4	9	1	3	2
2	4	8	3	6	1	5	7	9
1	9	3	5	7	2	6	8	4
9	3	5	6	2	8	7	4	1
8	1	2	4	5	7	3	9	6
4	6	7	9	1	3	8	2	5
5	8	1	2	3	4	9	6	7
3	7	4	1	9	6	2	5	8
6	2	9	7	8	5	4	1	3

Solution # 535

7	5	2	4	6	8	9	1	3
3	1	6	2	9	5	4	7	8
4	9	8	1	7	3	6	2	5
8	7	5	3	4	6	2	9	1
6	4	1	5	2	9	8	3	7
2	3	9	8	1	7	5	6	4
1	8	4	9	3	2	7	5	6
5	2	7	6	8	1	3	4	9
9	6	3	7	5	4	1	8	2

Solution # 536

8	6	7	1	2	5	9	4	3
9	4	1	7	3	6	8	5	2
3	2	5	9	8	4	6	1	7
5	3	4	6	9	8	7	2	1
7	9	6	4	1	2	5	3	8
1	8	2	5	7	3	4	9	6
6	5	8	3	4	1	2	7	9
2	7	3	8	5	9	1	6	4
4	1	9	2	6	7	3	8	5

Solution # 537

4	7	5	8	1	3	6	9	2
3	2	9	7	6	5	8	4	1
6	8	1	4	2	9	3	7	5
9	4	2	3	7	1	5	8	6
8	5	3	6	9	2	4	1	7
7	1	6	5	8	4	2	3	9
1	6	8	2	3	7	9	5	4
5	3	7	9	4	6	1	2	8
2	9	4	1	5	8	7	6	3

Solution # 538

1	3	5	2	4	8	9	7	6
7	8	4	5	6	9	1	2	3
2	6	9	1	7	3	5	4	8
3	5	7	6	2	1	4	8	9
8	4	1	9	5	7	3	6	2
9	2	6	3	8	4	7	1	5
4	1	2	8	3	5	6	9	7
5	7	8	4	9	6	2	3	1
6	9	3	7	1	2	8	5	4

Solution # 539

4	5	1	9	7	6	2	3	8
8	2	9	4	3	1	5	6	7
6	7	3	8	2	5	4	9	1
9	1	2	3	5	4	7	8	6
5	6	8	7	1	2	3	4	9
7	3	4	6	8	9	1	2	5
3	4	5	1	6	8	9	7	2
1	8	7	2	9	3	6	5	4
2	9	6	5	4	7	8	1	3

Solution # 540

7	3	5	2	9	8	6	1	4
8	6	2	3	1	4	7	9	5
9	1	4	5	6	7	2	8	3
4	9	6	8	5	2	1	3	7
5	2	3	1	7	6	8	4	9
1	7	8	4	3	9	5	6	2
3	4	7	6	2	1	9	5	8
2	5	1	9	8	3	4	7	6
6	8	9	7	4	5	3	2	1

Solution # 541

5	4	3	7	2	6	8	1	9
1	6	2	9	8	4	3	7	5
9	7	8	5	3	1	2	6	4
4	5	6	1	7	2	9	3	8
7	2	9	8	5	3	6	4	1
3	8	1	4	6	9	5	2	7
8	3	5	2	4	7	1	9	6
6	1	7	3	9	5	4	8	2
2	9	4	6	1	8	7	5	3

Solution # 542

3	4	9	5	8	2	7	6	1
8	2	7	1	6	4	3	9	5
1	6	5	3	7	9	8	2	4
5	7	2	4	9	3	1	8	6
4	9	1	8	5	6	2	7	3
6	8	3	2	1	7	4	5	9
2	5	8	9	3	1	6	4	7
9	3	6	7	4	8	5	1	2
7	1	4	6	2	5	9	3	8

Solution # 543

8	6	4	3	7	2	9	1	5
7	2	9	1	5	8	3	4	6
3	5	1	6	9	4	8	7	2
5	3	6	7	8	1	4	2	9
9	7	8	2	4	5	1	6	3
1	4	2	9	6	3	5	8	7
6	9	5	4	1	7	2	3	8
2	1	7	8	3	9	6	5	4
4	8	3	5	2	6	7	9	1

Solution # 544

7	9	1	8	3	6	2	4	5
4	6	8	7	2	5	1	3	9
3	5	2	4	9	1	6	8	7
5	4	7	2	6	9	8	1	3
9	1	6	3	8	4	5	7	2
2	8	3	1	5	7	4	9	6
1	2	9	5	4	3	7	6	8
8	3	4	6	7	2	9	5	1
6	7	5	9	1	8	3	2	4

Solution # 545

4	1	7	5	2	8	6	3	9
3	8	9	4	7	6	5	1	2
5	6	2	9	1	3	8	4	7
8	5	1	3	4	9	2	7	6
2	3	4	7	6	5	9	8	1
7	9	6	1	8	2	4	5	3
1	7	8	6	9	4	3	2	5
9	2	3	8	5	7	1	6	4
6	4	5	2	3	1	7	9	8

Solution # 546

5	3	1	8	7	6	4	2	9
9	2	6	3	5	4	8	1	7
4	8	7	2	9	1	5	6	3
1	4	8	6	3	5	9	7	2
6	9	5	1	2	7	3	8	4
3	7	2	4	8	9	1	5	6
7	1	4	5	6	3	2	9	8
2	6	3	9	1	8	7	4	5
8	5	9	7	4	2	6	3	1

Solution # 547

7	3	5	4	2	6	8	9	1
6	2	1	7	9	8	4	5	3
4	9	8	3	5	1	7	2	6
8	1	4	9	3	2	6	7	5
2	5	9	6	7	4	1	3	8
3	7	6	1	8	5	2	4	9
5	8	3	2	1	7	9	6	4
9	6	2	8	4	3	5	1	7
1	4	7	5	6	9	3	8	2

Solution # 548

6	8	3	5	2	9	1	7	4
4	1	9	8	7	6	3	2	5
5	7	2	4	1	3	9	6	8
9	6	4	7	3	8	2	5	1
3	5	8	1	4	2	6	9	7
7	2	1	6	9	5	4	8	3
8	3	5	2	6	4	7	1	9
1	9	6	3	8	7	5	4	2
2	4	7	9	5	1	8	3	6

Solution # 549

5	3	6	2	9	7	1	4	8
7	1	8	6	5	4	3	2	9
2	4	9	1	3	8	5	6	7
4	8	1	5	7	2	6	9	3
3	9	7	8	6	1	4	5	2
6	2	5	3	4	9	8	7	1
9	7	3	4	8	6	2	1	5
8	6	2	9	1	5	7	3	4
1	5	4	7	2	3	9	8	6

Solution # 550

3	8	5	9	7	2	4	1	6
2	4	6	3	8	1	9	7	5
9	1	7	6	5	4	8	3	2
6	5	9	4	1	7	3	2	8
8	3	2	5	6	9	1	4	7
1	7	4	8	2	3	6	5	9
5	9	1	2	4	6	7	8	3
4	6	8	7	3	5	2	9	1
7	2	3	1	9	8	5	6	4

Solution # 551

4	6	5	7	9	1	2	8	3
8	2	1	5	3	6	9	7	4
3	9	7	8	2	4	1	6	5
2	8	4	3	7	9	6	5	1
7	3	6	1	5	2	8	4	9
5	1	9	4	6	8	7	3	2
6	5	3	2	1	7	4	9	8
1	7	8	9	4	3	5	2	6
9	4	2	6	8	5	3	1	7

Solution # 552

8	4	9	5	3	6	2	1	7
5	6	7	2	4	1	3	8	9
3	2	1	7	8	9	5	6	4
4	3	6	8	1	7	9	5	2
1	7	5	9	6	2	8	4	3
2	9	8	4	5	3	1	7	6
9	8	2	1	7	4	6	3	5
6	1	4	3	9	5	7	2	8
7	5	3	6	2	8	4	9	1

Solution # 553

7	6	8	1	9	5	3	4	2
9	3	2	4	6	8	7	5	1
4	5	1	2	3	7	8	9	6
8	4	5	3	1	6	9	2	7
1	9	7	5	2	4	6	3	8
3	2	6	8	7	9	4	1	5
5	1	4	6	8	3	2	7	9
6	7	3	9	5	2	1	8	4
2	8	9	7	4	1	5	6	3

Solution # 554

4	6	2	5	9	7	1	3	8
8	9	1	3	2	4	7	6	5
3	5	7	8	6	1	2	4	9
2	4	9	1	5	3	6	8	7
6	3	5	2	7	8	9	1	4
7	1	8	9	4	6	5	2	3
5	2	6	4	8	9	3	7	1
1	7	4	6	3	5	8	9	2
9	8	3	7	1	2	4	5	6

Solution # 555

4	2	6	3	1	7	8	5	9
3	9	5	8	6	4	7	1	2
8	7	1	5	9	2	3	4	6
6	8	4	7	5	1	9	2	3
7	5	9	6	2	3	4	8	1
2	1	3	9	4	8	5	6	7
5	6	2	4	3	9	1	7	8
1	3	7	2	8	5	6	9	4
9	4	8	1	7	6	2	3	5

Solution # 556

3	5	6	7	1	9	4	2	8
9	2	8	3	6	4	1	7	5
7	4	1	5	2	8	9	3	6
6	9	5	1	8	7	3	4	2
2	7	3	4	9	5	8	6	1
1	8	4	2	3	6	7	5	9
5	1	7	8	4	2	6	9	3
4	3	9	6	5	1	2	8	7
8	6	2	9	7	3	5	1	4

Solution # 557

7	8	9	5	2	3	4	1	6
6	3	1	7	8	4	5	9	2
4	5	2	9	1	6	3	7	8
3	9	4	2	6	7	1	8	5
1	7	6	8	4	5	2	3	9
5	2	8	3	9	1	6	4	7
2	1	7	6	3	8	9	5	4
9	4	5	1	7	2	8	6	3
8	6	3	4	5	9	7	2	1

Solution # 558

3	5	2	9	6	4	7	1	8
8	6	1	7	5	3	4	9	2
7	4	9	2	1	8	3	5	6
1	2	4	8	9	6	5	3	7
6	7	3	4	2	5	1	8	9
5	9	8	1	3	7	6	2	4
2	1	7	5	4	9	8	6	3
9	8	6	3	7	1	2	4	5
4	3	5	6	8	2	9	7	1

Solution # 559

2	4	3	9	5	8	1	7	6
7	8	6	1	3	2	5	9	4
9	1	5	6	7	4	8	3	2
4	7	2	8	1	3	9	6	5
3	5	1	4	6	9	2	8	7
8	6	9	7	2	5	3	4	1
1	2	4	3	8	7	6	5	9
6	3	7	5	9	1	4	2	8
5	9	8	2	4	6	7	1	3

Solution # 560

7	3	9	4	5	8	2	6	1
6	5	8	7	1	2	4	3	9
2	4	1	3	9	6	8	5	7
9	2	7	1	3	5	6	8	4
5	6	4	8	7	9	1	2	3
8	1	3	6	2	4	9	7	5
1	8	5	2	4	3	7	9	6
3	7	6	9	8	1	5	4	2
4	9	2	5	6	7	3	1	8

Solution # 561

2	5	8	9	1	3	7	6	4
4	9	6	5	7	2	1	8	3
1	7	3	4	6	8	5	9	2
3	1	9	8	4	6	2	5	7
8	4	5	1	2	7	9	3	6
6	2	7	3	9	5	4	1	8
7	3	2	6	5	1	8	4	9
9	8	1	2	3	4	6	7	5
5	6	4	7	8	9	3	2	1

Solution # 562

3	9	5	1	7	6	2	4	8
7	4	8	3	5	2	1	6	9
6	2	1	4	8	9	5	3	7
9	6	4	5	2	7	3	8	1
5	3	2	8	4	1	9	7	6
8	1	7	6	9	3	4	5	2
1	7	3	2	6	4	8	9	5
2	5	9	7	3	8	6	1	4
4	8	6	9	1	5	7	2	3

Solution # 563

7	8	1	3	5	9	4	2	6
3	2	4	7	6	8	9	5	1
5	6	9	1	4	2	3	7	8
1	3	2	5	7	6	8	9	4
4	9	5	8	2	1	6	3	7
8	7	6	9	3	4	5	1	2
6	1	3	2	8	5	7	4	9
9	5	8	4	1	7	2	6	3
2	4	7	6	9	3	1	8	5

Solution # 564

2	3	9	6	1	4	5	8	7
6	7	8	9	5	2	3	1	4
5	4	1	7	8	3	6	2	9
7	2	6	3	4	1	9	5	8
1	9	4	8	7	5	2	3	6
8	5	3	2	6	9	4	7	1
3	6	7	5	9	8	1	4	2
9	1	2	4	3	7	8	6	5
4	8	5	1	2	6	7	9	3

Solution # 565

1	2	4	8	3	9	5	7	6
9	8	3	5	6	7	1	4	2
5	7	6	2	4	1	9	3	8
8	3	5	7	2	6	4	1	9
7	1	9	4	8	5	2	6	3
4	6	2	9	1	3	8	5	7
3	4	1	6	9	8	7	2	5
2	9	7	3	5	4	6	8	1
6	5	8	1	7	2	3	9	4

Solution # 566

4	2	6	3	5	9	1	8	7
3	1	5	6	8	7	4	9	2
7	8	9	4	2	1	3	5	6
5	4	7	9	3	8	6	2	1
6	3	1	2	4	5	8	7	9
2	9	8	7	1	6	5	4	3
9	7	3	8	6	4	2	1	5
1	6	4	5	7	2	9	3	8
8	5	2	1	9	3	7	6	4

Solution # 567

9	3	4	6	1	8	7	2	5
2	5	1	7	3	4	9	8	6
8	7	6	5	2	9	4	1	3
7	4	5	3	9	1	2	6	8
3	1	2	8	4	6	5	9	7
6	8	9	2	5	7	3	4	1
4	9	3	1	6	5	8	7	2
1	2	8	9	7	3	6	5	4
5	6	7	4	8	2	1	3	9

Solution # 568

9	5	1	3	2	8	4	6	7
2	6	4	5	7	9	8	1	3
8	7	3	4	1	6	2	9	5
3	1	6	7	8	2	5	4	9
5	4	2	9	3	1	6	7	8
7	9	8	6	5	4	3	2	1
4	8	5	2	9	7	1	3	6
6	3	9	1	4	5	7	8	2
1	2	7	8	6	3	9	5	4

Solution # 569

6	5	8	3	4	1	9	7	2
4	3	9	8	7	2	6	5	1
1	2	7	9	5	6	3	8	4
5	4	2	6	8	9	7	1	3
8	9	3	1	2	7	5	4	6
7	1	6	5	3	4	2	9	8
9	6	4	7	1	3	8	2	5
3	8	1	2	9	5	4	6	7
2	7	5	4	6	8	1	3	9

Solution # 570

3	5	2	8	9	1	6	4	7
7	1	9	5	6	4	2	8	3
4	8	6	3	2	7	9	1	5
6	7	4	9	5	8	1	3	2
1	3	5	2	4	6	8	7	9
2	9	8	7	1	3	5	6	4
9	4	7	1	8	5	3	2	6
8	2	3	6	7	9	4	5	1
5	6	1	4	3	2	7	9	8

Solution # 571

7	9	8	6	3	4	5	2	1
3	5	1	2	7	9	4	6	8
6	2	4	8	1	5	7	9	3
9	1	3	4	8	6	2	5	7
8	4	5	3	2	7	6	1	9
2	6	7	5	9	1	3	8	4
4	3	9	1	6	2	8	7	5
1	8	6	7	5	3	9	4	2
5	7	2	9	4	8	1	3	6

Solution # 572

6	8	7	5	3	9	1	2	4
3	2	1	7	8	4	6	9	5
9	4	5	2	1	6	3	8	7
2	7	4	3	9	5	8	6	1
8	6	3	1	4	2	5	7	9
5	1	9	8	6	7	4	3	2
4	3	8	9	7	1	2	5	6
1	9	2	6	5	3	7	4	8
7	5	6	4	2	8	9	1	3

Solution # 573

8	4	7	5	9	2	6	3	1
3	1	5	4	6	8	9	7	2
2	9	6	7	1	3	4	5	8
1	2	8	3	5	4	7	9	6
7	5	3	9	2	6	8	1	4
4	6	9	1	8	7	3	2	5
9	8	2	6	7	5	1	4	3
6	3	1	2	4	9	5	8	7
5	7	4	8	3	1	2	6	9

Solution # 574

4	5	7	6	3	9	2	1	8
2	3	1	4	7	8	9	5	6
9	8	6	1	2	5	4	3	7
7	4	3	5	9	6	1	8	2
6	1	2	7	8	4	5	9	3
5	9	8	3	1	2	6	7	4
8	7	4	9	6	1	3	2	5
1	2	5	8	4	3	7	6	9
3	6	9	2	5	7	8	4	1

Solution # 575

9	2	6	5	3	8	4	7	1
5	3	7	4	6	1	2	8	9
8	1	4	7	9	2	5	6	3
6	9	2	3	1	5	7	4	8
1	7	5	2	8	4	3	9	6
4	8	3	9	7	6	1	2	5
7	4	9	8	5	3	6	1	2
3	6	8	1	2	7	9	5	4
2	5	1	6	4	9	8	3	7

Solution # 576

5	3	4	1	9	2	6	7	8
2	1	8	7	5	6	9	3	4
9	7	6	3	8	4	1	2	5
8	2	5	9	6	3	4	1	7
3	6	1	2	4	7	5	8	9
7	4	9	5	1	8	3	6	2
1	9	2	8	3	5	7	4	6
4	5	7	6	2	1	8	9	3
6	8	3	4	7	9	2	5	1

Solution # 577

6	5	7	2	1	9	3	4	8
1	3	9	4	8	7	5	6	2
4	2	8	6	5	3	7	9	1
8	1	5	3	4	6	2	7	9
9	4	2	8	7	5	6	1	3
3	7	6	9	2	1	8	5	4
2	8	1	7	6	4	9	3	5
5	6	3	1	9	2	4	8	7
7	9	4	5	3	8	1	2	6

Solution # 578

2	3	6	8	4	7	5	1	9
1	4	7	2	5	9	3	6	8
5	9	8	3	1	6	7	2	4
4	6	2	5	9	8	1	3	7
9	7	5	1	6	3	4	8	2
3	8	1	7	2	4	6	9	5
6	5	4	9	3	2	8	7	1
8	1	9	6	7	5	2	4	3
7	2	3	4	8	1	9	5	6

Solution # 579

2	3	4	8	6	9	1	7	5
6	8	1	5	2	7	9	4	3
9	5	7	1	4	3	8	6	2
7	9	5	3	1	6	4	2	8
1	6	3	4	8	2	7	5	9
8	4	2	9	7	5	6	3	1
5	1	8	7	3	4	2	9	6
3	7	6	2	9	1	5	8	4
4	2	9	6	5	8	3	1	7

Solution # 580

6	2	1	9	5	8	4	7	3
7	5	9	3	4	2	1	8	6
3	4	8	6	1	7	5	2	9
5	3	4	1	7	9	2	6	8
1	7	6	8	2	3	9	5	4
9	8	2	5	6	4	3	1	7
8	9	5	2	3	6	7	4	1
2	6	7	4	9	1	8	3	5
4	1	3	7	8	5	6	9	2

Solution # 581

4	5	8	7	6	1	9	3	2
1	3	6	8	2	9	5	4	7
7	9	2	5	3	4	6	1	8
5	1	7	4	9	2	3	8	6
8	6	4	3	7	5	2	9	1
9	2	3	6	1	8	7	5	4
2	8	5	9	4	7	1	6	3
3	7	9	1	8	6	4	2	5
6	4	1	2	5	3	8	7	9

Solution # 582

7	2	1	9	3	5	6	4	8
5	6	8	4	1	2	7	3	9
9	3	4	7	8	6	5	1	2
2	1	6	3	5	7	8	9	4
8	4	7	1	6	9	2	5	3
3	5	9	8	2	4	1	6	7
6	9	5	2	4	8	3	7	1
1	7	2	5	9	3	4	8	6
4	8	3	6	7	1	9	2	5

Solution # 583

3	4	5	9	1	2	6	7	8
2	9	7	8	6	5	4	3	1
1	8	6	4	7	3	2	9	5
4	1	8	7	5	6	9	2	3
5	3	2	1	4	9	8	6	7
7	6	9	3	2	8	1	5	4
6	2	3	5	8	4	7	1	9
8	5	1	2	9	7	3	4	6
9	7	4	6	3	1	5	8	2

Solution # 584

7	8	3	5	2	4	6	9	1
2	5	9	1	6	7	4	3	8
6	1	4	3	9	8	5	7	2
5	3	6	7	1	9	2	8	4
8	9	1	4	3	2	7	6	5
4	2	7	6	8	5	3	1	9
3	6	8	2	4	1	9	5	7
9	7	2	8	5	3	1	4	6
1	4	5	9	7	6	8	2	3

Solution # 585

5	7	6	8	2	3	4	1	9
3	1	4	9	6	5	8	2	7
2	8	9	4	7	1	5	6	3
4	2	1	3	9	8	7	5	6
8	5	7	2	1	6	3	9	4
9	6	3	7	5	4	2	8	1
1	9	2	5	4	7	6	3	8
7	3	5	6	8	9	1	4	2
6	4	8	1	3	2	9	7	5

Solution # 586

3	5	4	8	1	6	7	9	2
7	2	8	4	3	9	1	5	6
6	1	9	7	2	5	3	8	4
2	4	7	6	5	1	9	3	8
9	3	5	2	8	4	6	7	1
1	8	6	3	9	7	4	2	5
8	9	1	5	6	3	2	4	7
5	7	3	1	4	2	8	6	9
4	6	2	9	7	8	5	1	3

Solution # 587

4	9	5	6	3	7	1	8	2
1	7	8	4	2	9	3	6	5
2	3	6	8	5	1	9	7	4
8	2	4	9	1	6	7	5	3
5	1	9	7	8	3	2	4	6
7	6	3	2	4	5	8	1	9
3	8	2	5	7	4	6	9	1
6	5	7	1	9	2	4	3	8
9	4	1	3	6	8	5	2	7

Solution # 588

8	5	6	3	2	9	4	7	1
7	2	3	1	4	5	6	8	9
4	1	9	8	6	7	5	3	2
1	7	5	4	8	2	9	6	3
6	4	2	7	9	3	1	5	8
3	9	8	6	5	1	2	4	7
9	6	7	2	3	4	8	1	5
2	3	4	5	1	8	7	9	6
5	8	1	9	7	6	3	2	4

Solution # 589

3	7	9	8	6	1	2	5	4
1	4	6	2	5	7	3	9	8
2	5	8	3	9	4	6	7	1
7	8	2	4	3	9	5	1	6
4	6	1	5	2	8	9	3	7
5	9	3	7	1	6	4	8	2
9	3	4	1	8	2	7	6	5
8	2	5	6	7	3	1	4	9
6	1	7	9	4	5	8	2	3

Solution # 590

1	7	8	4	6	2	3	5	9
2	5	3	7	8	9	6	1	4
4	6	9	1	5	3	7	2	8
8	3	1	6	9	4	5	7	2
7	2	4	8	1	5	9	6	3
5	9	6	2	3	7	4	8	1
9	8	5	3	2	6	1	4	7
6	1	7	9	4	8	2	3	5
3	4	2	5	7	1	8	9	6

Solution # 591

6	8	7	1	2	9	3	5	4
2	3	9	5	8	4	6	1	7
1	4	5	7	3	6	8	9	2
7	6	4	9	5	3	2	8	1
3	9	2	6	1	8	4	7	5
8	5	1	4	7	2	9	3	6
9	1	8	2	4	7	5	6	3
4	7	6	3	9	5	1	2	8
5	2	3	8	6	1	7	4	9

Solution # 592

6	4	3	9	5	7	2	1	8
9	8	5	1	3	2	7	6	4
2	1	7	8	4	6	9	3	5
4	3	1	7	8	5	6	9	2
5	6	9	3	2	1	8	4	7
8	7	2	4	6	9	3	5	1
1	9	8	6	7	4	5	2	3
7	2	4	5	9	3	1	8	6
3	5	6	2	1	8	4	7	9

Solution # 593

9	6	7	4	2	8	5	3	1
4	2	5	6	3	1	7	8	9
1	8	3	5	9	7	6	4	2
5	9	4	2	7	6	8	1	3
8	7	6	1	4	3	9	2	5
2	3	1	8	5	9	4	7	6
3	4	9	7	1	5	2	6	8
7	1	8	9	6	2	3	5	4
6	5	2	3	8	4	1	9	7

Solution # 594

4	8	3	2	5	6	9	7	1
2	5	1	3	7	9	8	6	4
7	6	9	4	8	1	5	2	3
5	3	6	9	4	7	2	1	8
1	9	7	8	3	2	4	5	6
8	4	2	1	6	5	3	9	7
9	7	4	5	1	3	6	8	2
3	1	5	6	2	8	7	4	9
6	2	8	7	9	4	1	3	5

Solution # 595

4	9	8	1	2	5	7	3	6
2	6	3	7	8	9	5	4	1
7	5	1	4	3	6	2	8	9
6	3	7	5	1	4	9	2	8
1	4	9	2	6	8	3	7	5
5	8	2	3	9	7	6	1	4
3	2	4	6	5	1	8	9	7
8	1	6	9	7	2	4	5	3
9	7	5	8	4	3	1	6	2

Solution # 596

5	8	1	2	6	4	7	3	9
9	6	2	1	7	3	5	8	4
4	3	7	8	5	9	2	1	6
7	5	8	9	3	6	4	2	1
3	9	4	5	2	1	6	7	8
2	1	6	7	4	8	9	5	3
6	4	5	3	1	2	8	9	7
1	2	9	6	8	7	3	4	5
8	7	3	4	9	5	1	6	2

Solution # 597

8	6	2	5	1	7	4	3	9
4	9	5	2	3	6	7	8	1
3	1	7	9	4	8	2	5	6
9	8	3	6	7	2	1	4	5
5	4	6	1	9	3	8	7	2
7	2	1	8	5	4	9	6	3
2	5	4	3	8	9	6	1	7
6	3	8	7	2	1	5	9	4
1	7	9	4	6	5	3	2	8

Solution # 598

7	4	1	5	2	6	9	8	3
5	9	8	1	7	3	4	6	2
6	3	2	9	4	8	1	7	5
4	7	9	8	1	2	5	3	6
8	1	3	6	5	4	7	2	9
2	5	6	3	9	7	8	1	4
3	2	7	4	8	5	6	9	1
9	8	5	2	6	1	3	4	7
1	6	4	7	3	9	2	5	8

Solution # 599

3	8	4	9	5	2	7	6	1
5	6	9	4	7	1	8	3	2
2	1	7	3	8	6	4	5	9
9	5	2	6	3	4	1	8	7
7	4	8	1	2	5	6	9	3
6	3	1	7	9	8	2	4	5
1	7	5	8	4	9	3	2	6
8	2	6	5	1	3	9	7	4
4	9	3	2	6	7	5	1	8

Solution # 600

9	3	6	5	4	8	7	2	1
7	8	5	3	1	2	9	6	4
4	2	1	7	6	9	5	8	3
1	6	9	4	2	5	8	3	7
2	5	7	8	3	1	6	4	9
8	4	3	6	9	7	2	1	5
6	1	2	9	7	3	4	5	8
3	9	8	2	5	4	1	7	6
5	7	4	1	8	6	3	9	2

Solution # 601

8	1	3	4	7	5	6	9	2
4	2	9	1	6	3	8	5	7
5	6	7	8	9	2	1	4	3
3	4	5	6	2	8	7	1	9
6	8	1	9	4	7	2	3	5
7	9	2	3	5	1	4	8	6
1	5	6	7	8	9	3	2	4
2	7	8	5	3	4	9	6	1
9	3	4	2	1	6	5	7	8

Solution # 602

7	1	3	5	8	4	9	2	6
5	8	6	7	2	9	3	4	1
4	9	2	1	3	6	8	7	5
9	5	1	4	7	8	6	3	2
3	6	4	9	1	2	7	5	8
8	2	7	3	6	5	4	1	9
6	7	8	2	4	1	5	9	3
1	4	5	6	9	3	2	8	7
2	3	9	8	5	7	1	6	4

Solution # 603

7	3	8	5	6	4	2	1	9
2	4	9	3	8	1	7	6	5
1	6	5	2	7	9	3	4	8
3	2	4	7	1	5	9	8	6
8	1	7	6	9	3	4	5	2
5	9	6	8	4	2	1	3	7
4	8	3	9	2	6	5	7	1
9	7	1	4	5	8	6	2	3
6	5	2	1	3	7	8	9	4

Solution # 604

5	7	6	1	4	3	9	8	2
3	2	8	9	7	6	1	5	4
1	9	4	5	2	8	7	3	6
4	5	7	2	6	9	3	1	8
8	6	1	4	3	7	2	9	5
2	3	9	8	5	1	6	4	7
7	4	3	6	9	5	8	2	1
6	8	5	3	1	2	4	7	9
9	1	2	7	8	4	5	6	3

Solution # 605

1	9	4	6	8	2	3	7	5
3	5	7	1	4	9	2	6	8
6	2	8	5	3	7	4	1	9
2	1	6	4	5	8	7	9	3
8	7	9	2	1	3	5	4	6
5	4	3	7	9	6	1	8	2
7	8	2	3	6	4	9	5	1
4	6	1	9	2	5	8	3	7
9	3	5	8	7	1	6	2	4

Solution # 606

7	8	2	9	6	3	1	4	5
6	3	4	1	5	7	9	8	2
5	1	9	2	8	4	3	7	6
2	6	7	3	1	9	4	5	8
3	4	8	5	7	2	6	1	9
1	9	5	8	4	6	7	2	3
9	7	1	6	2	5	8	3	4
8	5	6	4	3	1	2	9	7
4	2	3	7	9	8	5	6	1

Solution # 607

4	6	8	7	3	2	9	1	5
3	9	2	4	5	1	6	8	7
7	5	1	6	8	9	4	3	2
5	3	6	8	9	4	7	2	1
2	1	4	3	6	7	8	5	9
8	7	9	1	2	5	3	6	4
1	8	5	9	4	3	2	7	6
9	2	3	5	7	6	1	4	8
6	4	7	2	1	8	5	9	3

Solution # 608

2	3	8	1	5	7	4	9	6
4	1	9	8	6	2	7	5	3
5	6	7	4	3	9	2	1	8
3	8	5	2	4	1	9	6	7
9	4	1	6	7	3	8	2	5
7	2	6	5	9	8	3	4	1
6	7	4	3	2	5	1	8	9
8	9	2	7	1	6	5	3	4
1	5	3	9	8	4	6	7	2

Solution # 609

3	7	9	4	5	2	8	6	1
8	4	2	7	6	1	3	9	5
5	1	6	3	8	9	4	2	7
6	2	5	1	4	8	7	3	9
1	3	7	6	9	5	2	8	4
4	9	8	2	7	3	5	1	6
7	8	1	5	2	6	9	4	3
9	6	4	8	3	7	1	5	2
2	5	3	9	1	4	6	7	8

Solution # 610

7	1	6	5	9	4	2	8	3
5	3	4	2	8	7	1	9	6
2	9	8	1	6	3	7	4	5
9	4	1	3	2	8	6	5	7
3	7	5	6	4	1	9	2	8
8	6	2	9	7	5	3	1	4
4	2	9	8	3	6	5	7	1
1	8	3	7	5	2	4	6	9
6	5	7	4	1	9	8	3	2

Solution # 611

8	3	5	4	7	9	2	6	1
7	9	6	2	8	1	4	3	5
4	2	1	3	6	5	8	7	9
2	4	8	5	9	6	7	1	3
3	6	7	1	2	8	5	9	4
1	5	9	7	3	4	6	2	8
6	1	3	8	5	7	9	4	2
5	7	2	9	4	3	1	8	6
9	8	4	6	1	2	3	5	7

Solution # 612

4	3	9	6	5	8	7	1	2
6	5	1	9	7	2	4	8	3
2	7	8	1	3	4	5	6	9
1	9	2	5	8	3	6	7	4
5	8	3	4	6	7	9	2	1
7	6	4	2	1	9	8	3	5
8	2	5	3	4	6	1	9	7
3	4	6	7	9	1	2	5	8
9	1	7	8	2	5	3	4	6

Solution # 613

3	5	9	2	6	8	4	1	7
2	8	1	9	7	4	6	5	3
7	4	6	5	3	1	2	9	8
6	3	5	1	9	2	8	7	4
4	9	7	6	8	3	1	2	5
8	1	2	7	4	5	3	6	9
9	7	8	4	2	6	5	3	1
1	6	3	8	5	7	9	4	2
5	2	4	3	1	9	7	8	6

Solution # 614

4	5	3	2	6	7	1	8	9
2	8	9	1	5	3	6	4	7
6	7	1	9	8	4	5	2	3
5	2	4	3	9	6	8	7	1
7	3	8	4	2	1	9	5	6
1	9	6	8	7	5	2	3	4
8	6	2	7	3	9	4	1	5
9	4	7	5	1	8	3	6	2
3	1	5	6	4	2	7	9	8

Solution # 615

7	4	3	6	5	8	1	2	9
1	5	2	7	3	9	4	6	8
8	9	6	1	2	4	3	7	5
3	1	7	8	9	6	5	4	2
9	6	4	2	1	5	8	3	7
2	8	5	3	4	7	9	1	6
4	3	9	5	7	2	6	8	1
6	2	1	9	8	3	7	5	4
5	7	8	4	6	1	2	9	3

Solution # 616

6	2	4	9	1	3	7	5	8
8	1	3	5	6	7	9	2	4
5	9	7	4	2	8	3	6	1
2	6	1	7	8	4	5	9	3
4	8	9	2	3	5	6	1	7
7	3	5	1	9	6	4	8	2
9	7	2	6	4	1	8	3	5
3	4	6	8	5	2	1	7	9
1	5	8	3	7	9	2	4	6

Solution # 617

5	3	7	2	6	4	8	9	1
6	4	2	1	9	8	5	3	7
8	1	9	3	7	5	4	6	2
9	6	4	5	3	7	2	1	8
3	2	8	4	1	9	7	5	6
7	5	1	8	2	6	9	4	3
2	9	6	7	4	1	3	8	5
4	8	3	6	5	2	1	7	9
1	7	5	9	8	3	6	2	4

Solution # 618

6	8	3	4	7	5	2	1	9
5	4	2	8	1	9	7	3	6
9	7	1	2	6	3	5	4	8
7	2	5	3	4	8	6	9	1
4	3	9	6	5	1	8	7	2
8	1	6	9	2	7	3	5	4
1	6	4	7	3	2	9	8	5
2	9	7	5	8	4	1	6	3
3	5	8	1	9	6	4	2	7

Solution # 619

9	6	4	2	8	5	7	3	1
5	3	1	4	9	7	8	6	2
7	8	2	6	3	1	5	4	9
4	7	6	3	5	2	9	1	8
3	5	9	8	1	4	2	7	6
1	2	8	9	7	6	3	5	4
8	4	5	7	6	9	1	2	3
6	9	7	1	2	3	4	8	5
2	1	3	5	4	8	6	9	7

Solution # 620

3	1	8	4	2	9	7	5	6
5	7	4	8	3	6	9	2	1
6	2	9	5	1	7	4	8	3
7	4	1	6	8	5	3	9	2
2	8	6	3	9	1	5	7	4
9	5	3	7	4	2	1	6	8
4	3	2	9	7	8	6	1	5
1	6	7	2	5	4	8	3	9
8	9	5	1	6	3	2	4	7

Solution # 621

4	3	7	5	6	1	9	2	8
8	5	2	4	3	9	1	6	7
9	6	1	7	2	8	5	4	3
5	9	4	1	8	6	7	3	2
2	1	3	9	4	7	8	5	6
6	7	8	2	5	3	4	9	1
7	4	6	8	9	2	3	1	5
1	2	5	3	7	4	6	8	9
3	8	9	6	1	5	2	7	4

Solution # 622

8	7	3	1	6	2	5	9	4
1	6	5	4	7	9	8	3	2
9	2	4	5	3	8	6	7	1
2	3	8	7	5	1	4	6	9
7	4	9	3	8	6	1	2	5
6	5	1	9	2	4	7	8	3
4	1	6	2	9	7	3	5	8
5	9	7	8	1	3	2	4	6
3	8	2	6	4	5	9	1	7

Solution # 623

1	4	3	7	9	5	6	8	2
7	8	6	2	3	1	5	4	9
5	2	9	4	6	8	7	3	1
6	1	2	9	4	3	8	7	5
3	5	7	8	2	6	9	1	4
4	9	8	1	5	7	3	2	6
9	3	4	6	8	2	1	5	7
2	7	5	3	1	9	4	6	8
8	6	1	5	7	4	2	9	3

Solution # 624

3	4	8	5	1	2	9	6	7
2	9	7	4	6	8	5	3	1
6	1	5	3	7	9	8	4	2
5	3	6	2	4	7	1	9	8
9	2	4	8	3	1	7	5	6
7	8	1	9	5	6	3	2	4
8	6	9	7	2	5	4	1	3
1	7	3	6	9	4	2	8	5
4	5	2	1	8	3	6	7	9

Solution # 625

7	2	1	3	8	5	9	6	4
5	9	8	1	6	4	2	3	7
3	6	4	7	2	9	8	1	5
8	1	5	4	9	3	6	7	2
6	4	9	2	5	7	1	8	3
2	3	7	8	1	6	5	4	9
4	5	2	6	3	8	7	9	1
9	8	3	5	7	1	4	2	6
1	7	6	9	4	2	3	5	8

Solution # 626

3	9	8	2	7	6	4	5	1
1	2	5	4	9	3	6	7	8
7	6	4	8	1	5	3	2	9
2	7	6	1	4	8	9	3	5
4	5	9	7	3	2	8	1	6
8	3	1	5	6	9	2	4	7
9	4	7	6	2	1	5	8	3
5	1	3	9	8	4	7	6	2
6	8	2	3	5	7	1	9	4

Solution # 627

4	3	8	2	5	7	9	6	1
9	2	1	3	6	4	5	8	7
7	6	5	8	1	9	4	3	2
6	8	2	4	7	1	3	5	9
5	1	4	9	8	3	2	7	6
3	7	9	6	2	5	1	4	8
8	4	6	1	3	2	7	9	5
2	9	7	5	4	8	6	1	3
1	5	3	7	9	6	8	2	4

Solution # 628

1	7	3	5	2	6	9	4	8
8	6	9	3	4	1	2	7	5
4	5	2	9	8	7	3	1	6
3	9	8	6	7	5	1	2	4
2	1	6	4	9	3	5	8	7
5	4	7	2	1	8	6	9	3
9	3	4	7	5	2	8	6	1
6	2	1	8	3	4	7	5	9
7	8	5	1	6	9	4	3	2

Solution # 629

5	6	2	3	8	1	4	7	9
8	1	9	4	7	2	6	5	3
4	3	7	6	9	5	8	2	1
3	4	5	2	6	7	1	9	8
2	8	1	5	4	9	3	6	7
9	7	6	1	3	8	2	4	5
7	9	4	8	1	6	5	3	2
6	2	8	7	5	3	9	1	4
1	5	3	9	2	4	7	8	6

Solution # 630

5	6	9	1	4	3	2	7	8
1	3	2	7	8	9	5	6	4
7	4	8	5	2	6	3	1	9
9	8	6	2	7	5	4	3	1
4	2	5	3	9	1	7	8	6
3	7	1	4	6	8	9	5	2
6	9	3	8	5	2	1	4	7
2	1	4	6	3	7	8	9	5
8	5	7	9	1	4	6	2	3

Solution # 631

8	4	5	6	9	7	3	2	1
2	3	7	4	5	1	9	6	8
9	6	1	8	2	3	4	5	7
7	1	8	3	4	5	2	9	6
4	2	3	7	6	9	1	8	5
6	5	9	2	1	8	7	4	3
5	9	4	1	3	6	8	7	2
3	7	6	9	8	2	5	1	4
1	8	2	5	7	4	6	3	9

Solution # 632

1	6	4	3	5	2	7	8	9
5	7	8	9	1	4	6	2	3
9	3	2	6	7	8	5	1	4
3	1	7	2	4	5	8	9	6
2	5	6	1	8	9	3	4	7
4	8	9	7	6	3	1	5	2
7	4	1	5	2	6	9	3	8
8	9	5	4	3	7	2	6	1
6	2	3	8	9	1	4	7	5

Solution # 633

5	2	3	9	7	8	4	1	6
1	6	9	4	5	3	2	7	8
4	8	7	2	1	6	3	9	5
9	1	8	3	6	5	7	2	4
6	5	4	7	9	2	1	8	3
7	3	2	1	8	4	5	6	9
3	9	1	8	4	7	6	5	2
8	4	5	6	2	1	9	3	7
2	7	6	5	3	9	8	4	1

Solution # 634

3	2	1	4	9	8	5	6	7
9	6	5	2	7	1	8	4	3
8	4	7	6	3	5	9	1	2
5	3	6	1	4	7	2	9	8
1	9	2	8	5	6	3	7	4
7	8	4	9	2	3	1	5	6
4	1	8	3	6	9	7	2	5
6	7	3	5	1	2	4	8	9
2	5	9	7	8	4	6	3	1

Solution # 635

6	7	4	9	2	1	5	3	8
8	1	3	6	4	5	9	2	7
5	2	9	3	8	7	6	1	4
7	3	5	4	1	6	8	9	2
9	4	2	8	5	3	1	7	6
1	6	8	2	7	9	4	5	3
2	5	6	1	3	8	7	4	9
4	8	1	7	9	2	3	6	5
3	9	7	5	6	4	2	8	1

Solution # 636

9	2	4	8	7	1	5	3	6
8	7	5	3	4	6	1	9	2
6	1	3	9	5	2	4	7	8
7	4	9	1	6	3	2	8	5
1	5	6	7	2	8	9	4	3
2	3	8	5	9	4	6	1	7
5	8	7	6	1	9	3	2	4
3	9	2	4	8	5	7	6	1
4	6	1	2	3	7	8	5	9

Solution # 637

2	3	7	9	8	5	4	6	1
1	5	9	7	4	6	3	8	2
4	8	6	1	3	2	7	5	9
9	7	5	4	1	8	6	2	3
8	6	1	3	2	7	5	9	4
3	2	4	5	6	9	8	1	7
5	1	3	6	9	4	2	7	8
7	9	8	2	5	3	1	4	6
6	4	2	8	7	1	9	3	5

Solution # 638

8	3	9	2	4	1	6	5	7
4	6	2	8	7	5	9	1	3
1	7	5	9	3	6	2	4	8
9	4	3	1	6	7	5	8	2
7	2	8	3	5	9	1	6	4
6	5	1	4	2	8	3	7	9
5	9	7	6	8	2	4	3	1
2	8	4	5	1	3	7	9	6
3	1	6	7	9	4	8	2	5

Solution # 639

9	3	6	1	8	5	4	7	2
4	7	5	6	2	9	8	1	3
2	8	1	7	4	3	6	9	5
3	9	8	4	6	7	5	2	1
7	1	4	8	5	2	3	6	9
5	6	2	9	3	1	7	8	4
1	4	9	3	7	6	2	5	8
6	2	3	5	1	8	9	4	7
8	5	7	2	9	4	1	3	6

Solution # 640

2	8	6	5	9	7	4	1	3
5	9	1	3	8	4	2	7	6
3	4	7	1	6	2	8	9	5
6	5	4	7	1	8	9	3	2
9	7	2	4	5	3	1	6	8
8	1	3	6	2	9	5	4	7
7	6	5	8	4	1	3	2	9
4	2	8	9	3	6	7	5	1
1	3	9	2	7	5	6	8	4

Solution # 641

4	6	2	7	1	8	9	5	3
3	1	9	4	5	2	6	7	8
8	7	5	3	6	9	4	2	1
7	4	6	8	2	1	5	3	9
2	3	8	9	4	5	1	6	7
9	5	1	6	7	3	8	4	2
6	8	4	2	9	7	3	1	5
5	9	7	1	3	6	2	8	4
1	2	3	5	8	4	7	9	6

Solution # 642

5	6	8	2	7	3	9	4	1
4	9	2	8	1	6	7	5	3
7	3	1	4	5	9	2	6	8
8	5	7	6	3	4	1	2	9
3	2	4	1	9	5	6	8	7
6	1	9	7	8	2	4	3	5
2	8	5	9	6	1	3	7	4
1	4	3	5	2	7	8	9	6
9	7	6	3	4	8	5	1	2

Solution # 643

1	6	5	9	3	8	2	4	7
3	9	7	4	2	1	5	6	8
4	8	2	5	7	6	1	3	9
7	1	8	3	9	4	6	2	5
9	4	6	8	5	2	3	7	1
2	5	3	6	1	7	8	9	4
6	2	9	1	4	5	7	8	3
5	7	4	2	8	3	9	1	6
8	3	1	7	6	9	4	5	2

Solution # 644

4	5	3	1	9	7	2	6	8
8	7	2	4	6	5	9	1	3
6	1	9	8	2	3	4	5	7
1	8	7	6	5	2	3	4	9
2	4	5	9	3	8	1	7	6
3	9	6	7	1	4	5	8	2
9	6	4	2	8	1	7	3	5
7	3	8	5	4	9	6	2	1
5	2	1	3	7	6	8	9	4

Solution # 645

3	7	2	1	9	8	5	4	6
1	8	5	3	6	4	2	7	9
9	4	6	2	7	5	1	3	8
5	1	8	9	4	7	6	2	3
7	6	4	8	3	2	9	1	5
2	9	3	6	5	1	7	8	4
8	3	7	5	1	6	4	9	2
4	5	9	7	2	3	8	6	1
6	2	1	4	8	9	3	5	7

Solution # 646

5	6	4	9	1	3	7	2	8
7	2	3	4	5	8	6	1	9
9	1	8	6	7	2	3	5	4
8	3	1	5	9	6	4	7	2
6	9	2	8	4	7	5	3	1
4	5	7	2	3	1	9	8	6
1	4	9	3	8	5	2	6	7
3	7	6	1	2	9	8	4	5
2	8	5	7	6	4	1	9	3

Solution # 647

1	5	7	9	4	6	3	2	8
4	8	9	2	5	3	1	7	6
3	2	6	7	8	1	5	4	9
9	1	2	3	6	5	7	8	4
6	4	3	8	9	7	2	5	1
8	7	5	1	2	4	9	6	3
7	6	1	4	3	2	8	9	5
2	9	4	5	1	8	6	3	7
5	3	8	6	7	9	4	1	2

Solution # 648

7	6	3	2	5	9	1	8	4
9	8	1	6	7	4	3	5	2
5	2	4	3	8	1	6	7	9
3	5	9	4	6	8	2	1	7
8	7	2	1	3	5	4	9	6
4	1	6	7	9	2	8	3	5
2	9	5	8	1	6	7	4	3
6	3	8	9	4	7	5	2	1
1	4	7	5	2	3	9	6	8

Solution # 649

7	3	4	9	8	1	5	2	6
8	9	5	6	7	2	3	4	1
2	1	6	4	5	3	8	7	9
1	2	9	5	4	7	6	3	8
5	4	3	2	6	8	9	1	7
6	7	8	1	3	9	2	5	4
3	5	1	7	9	6	4	8	2
4	6	2	8	1	5	7	9	3
9	8	7	3	2	4	1	6	5

Solution # 650

4	7	8	3	2	1	5	6	9
5	3	1	9	4	6	8	7	2
6	9	2	5	8	7	1	4	3
7	5	6	1	3	9	4	2	8
2	4	9	6	7	8	3	1	5
8	1	3	4	5	2	7	9	6
3	2	5	7	6	4	9	8	1
1	8	7	2	9	3	6	5	4
9	6	4	8	1	5	2	3	7

Solution # 651

6	3	8	7	1	9	2	5	4
4	1	2	5	8	3	9	6	7
5	7	9	2	6	4	8	3	1
8	5	1	3	4	2	6	7	9
3	9	7	6	5	8	4	1	2
2	4	6	1	9	7	5	8	3
1	2	5	9	7	6	3	4	8
9	6	4	8	3	1	7	2	5
7	8	3	4	2	5	1	9	6

Solution # 652

5	2	3	7	4	8	6	1	9
4	9	8	6	1	3	7	2	5
6	7	1	5	2	9	4	3	8
3	8	7	4	6	1	9	5	2
1	5	6	9	7	2	8	4	3
2	4	9	3	8	5	1	6	7
9	6	2	8	5	4	3	7	1
8	1	4	2	3	7	5	9	6
7	3	5	1	9	6	2	8	4

Solution # 653

7	4	9	1	8	5	2	3	6
2	5	6	3	9	4	1	8	7
1	8	3	6	7	2	5	9	4
6	1	2	7	3	9	8	4	5
4	7	8	2	5	1	3	6	9
3	9	5	8	4	6	7	1	2
9	3	1	5	6	7	4	2	8
5	2	4	9	1	8	6	7	3
8	6	7	4	2	3	9	5	1

Solution # 654

4	6	3	8	9	1	7	2	5
1	2	9	3	5	7	4	6	8
7	8	5	6	2	4	3	9	1
6	3	4	1	7	2	5	8	9
2	5	7	9	3	8	6	1	4
9	1	8	4	6	5	2	3	7
8	9	6	7	4	3	1	5	2
3	7	2	5	1	9	8	4	6
5	4	1	2	8	6	9	7	3

Solution # 655

8	5	6	7	1	3	4	9	2
3	2	7	9	8	4	1	5	6
4	1	9	6	2	5	7	3	8
6	4	1	2	3	9	5	8	7
9	8	5	4	6	7	3	2	1
7	3	2	8	5	1	6	4	9
1	9	3	5	7	2	8	6	4
5	6	4	1	9	8	2	7	3
2	7	8	3	4	6	9	1	5

Solution # 656

9	8	6	7	2	3	4	5	1
2	1	7	8	5	4	6	9	3
5	4	3	9	1	6	8	7	2
7	2	8	3	4	9	5	1	6
3	5	1	2	6	8	7	4	9
6	9	4	5	7	1	3	2	8
4	3	9	1	8	5	2	6	7
8	7	5	6	9	2	1	3	4
1	6	2	4	3	7	9	8	5

Solution # 657

5	8	3	4	2	9	7	6	1
1	2	6	7	5	8	3	9	4
4	9	7	6	1	3	2	5	8
7	5	2	9	3	1	4	8	6
9	3	4	5	8	6	1	7	2
8	6	1	2	4	7	9	3	5
3	4	5	8	9	2	6	1	7
6	1	8	3	7	4	5	2	9
2	7	9	1	6	5	8	4	3

Solution # 658

8	7	1	2	4	5	3	9	6
5	9	4	7	3	6	2	8	1
6	2	3	8	1	9	7	5	4
4	6	9	3	5	2	8	1	7
2	5	7	1	9	8	4	6	3
3	1	8	4	6	7	5	2	9
9	8	2	6	7	4	1	3	5
7	3	5	9	2	1	6	4	8
1	4	6	5	8	3	9	7	2

Solution # 659

2	6	4	7	1	5	8	9	3
5	9	7	3	8	2	6	1	4
8	1	3	9	6	4	5	7	2
3	5	9	1	7	8	2	4	6
6	7	2	4	5	3	9	8	1
1	4	8	6	2	9	3	5	7
9	2	6	8	4	7	1	3	5
4	3	5	2	9	1	7	6	8
7	8	1	5	3	6	4	2	9

Solution # 660

5	9	2	4	3	6	1	7	8
4	7	3	1	2	8	5	6	9
8	6	1	5	7	9	4	2	3
1	5	9	3	4	2	6	8	7
7	4	6	9	8	5	2	3	1
3	2	8	6	1	7	9	5	4
9	8	4	2	6	3	7	1	5
6	3	5	7	9	1	8	4	2
2	1	7	8	5	4	3	9	6

Solution # 661

9	7	6	8	2	4	3	5	1
4	1	2	5	3	6	9	7	8
3	8	5	1	9	7	6	2	4
5	4	3	9	7	8	2	1	6
8	6	1	2	4	3	5	9	7
7	2	9	6	1	5	4	8	3
6	5	4	7	8	9	1	3	2
2	9	8	3	6	1	7	4	5
1	3	7	4	5	2	8	6	9

Solution # 662

7	2	4	6	5	9	3	1	8
3	9	8	1	2	7	5	4	6
5	6	1	4	3	8	7	9	2
8	7	6	5	9	4	1	2	3
2	5	3	8	6	1	9	7	4
4	1	9	3	7	2	6	8	5
9	8	5	2	1	3	4	6	7
1	3	2	7	4	6	8	5	9
6	4	7	9	8	5	2	3	1

Solution # 663

5	7	2	1	3	9	4	6	8
1	6	9	4	8	7	3	5	2
3	8	4	2	5	6	7	9	1
4	1	5	9	7	2	8	3	6
6	9	8	3	4	5	2	1	7
7	2	3	8	6	1	5	4	9
8	5	7	6	1	3	9	2	4
2	4	6	5	9	8	1	7	3
9	3	1	7	2	4	6	8	5

Solution # 664

5	8	3	9	4	6	2	7	1
4	2	6	5	7	1	9	3	8
9	1	7	3	8	2	5	6	4
3	5	2	8	6	7	1	4	9
8	6	4	1	5	9	3	2	7
1	7	9	2	3	4	8	5	6
7	3	8	6	1	5	4	9	2
2	4	1	7	9	3	6	8	5
6	9	5	4	2	8	7	1	3

Solution # 665

9	3	5	2	8	7	4	1	6
2	1	8	6	4	5	7	3	9
6	4	7	9	1	3	2	5	8
7	5	9	4	6	8	3	2	1
3	6	2	5	9	1	8	4	7
4	8	1	7	3	2	9	6	5
1	2	4	8	7	6	5	9	3
8	9	3	1	5	4	6	7	2
5	7	6	3	2	9	1	8	4

Solution # 666

5	9	3	2	7	8	6	4	1
4	7	6	5	1	3	8	9	2
1	8	2	4	6	9	5	7	3
6	5	1	9	4	2	3	8	7
7	3	9	1	8	5	2	6	4
2	4	8	7	3	6	9	1	5
3	1	5	8	9	4	7	2	6
8	2	4	6	5	7	1	3	9
9	6	7	3	2	1	4	5	8

Solution # 667

7	4	8	3	5	9	6	2	1
1	9	3	8	6	2	4	7	5
2	6	5	7	1	4	8	9	3
8	7	6	1	9	3	2	5	4
3	1	9	2	4	5	7	6	8
4	5	2	6	7	8	1	3	9
6	3	1	5	8	7	9	4	2
9	2	7	4	3	1	5	8	6
5	8	4	9	2	6	3	1	7

Solution # 668

1	4	9	6	8	7	2	5	3
2	3	6	5	9	1	4	8	7
8	5	7	3	2	4	9	6	1
4	6	8	1	3	2	7	9	5
7	1	2	9	6	5	3	4	8
3	9	5	4	7	8	1	2	6
5	2	4	8	1	3	6	7	9
6	7	3	2	5	9	8	1	4
9	8	1	7	4	6	5	3	2

Solution # 669

9	8	1	5	3	2	6	4	7
6	2	7	4	8	1	9	5	3
3	5	4	6	7	9	2	8	1
8	7	5	3	9	4	1	6	2
2	9	3	8	1	6	4	7	5
4	1	6	2	5	7	3	9	8
1	4	2	7	6	5	8	3	9
5	3	9	1	4	8	7	2	6
7	6	8	9	2	3	5	1	4

Solution # 670

4	9	6	8	5	2	3	7	1
1	8	3	6	7	9	5	4	2
7	5	2	1	3	4	8	6	9
3	6	4	2	1	7	9	5	8
8	1	7	4	9	5	2	3	6
9	2	5	3	6	8	4	1	7
2	3	8	7	4	1	6	9	5
5	4	1	9	8	6	7	2	3
6	7	9	5	2	3	1	8	4

Solution # 671

5	1	7	9	2	8	3	4	6
2	8	3	4	6	1	5	9	7
9	4	6	5	7	3	8	1	2
8	5	4	3	1	2	7	6	9
7	3	2	6	4	9	1	5	8
1	6	9	8	5	7	4	2	3
4	2	1	7	3	6	9	8	5
3	9	5	2	8	4	6	7	1
6	7	8	1	9	5	2	3	4

Solution # 672

6	9	3	7	8	1	2	5	4
2	4	8	9	5	3	6	7	1
5	7	1	2	6	4	3	9	8
8	6	4	5	7	9	1	2	3
9	3	7	1	2	8	5	4	6
1	5	2	4	3	6	7	8	9
7	1	9	6	4	2	8	3	5
3	2	6	8	9	5	4	1	7
4	8	5	3	1	7	9	6	2

Solution # 673

7	8	1	2	9	6	4	5	3
5	2	9	1	4	3	6	7	8
6	3	4	7	8	5	1	9	2
4	9	6	5	7	8	3	2	1
1	7	8	3	2	4	5	6	9
3	5	2	6	1	9	8	4	7
2	1	3	4	5	7	9	8	6
9	4	7	8	6	1	2	3	5
8	6	5	9	3	2	7	1	4

Solution # 674

2	4	9	1	7	5	3	6	8
1	3	6	8	4	2	5	9	7
5	7	8	9	3	6	2	4	1
7	8	4	6	1	3	9	2	5
6	9	2	5	8	7	1	3	4
3	1	5	2	9	4	8	7	6
8	6	1	4	2	9	7	5	3
9	5	7	3	6	1	4	8	2
4	2	3	7	5	8	6	1	9

Solution # 675

1	7	9	8	6	2	5	4	3
4	6	2	7	3	5	8	1	9
8	5	3	9	1	4	7	6	2
7	4	6	5	9	3	1	2	8
3	1	8	2	7	6	4	9	5
9	2	5	4	8	1	3	7	6
5	8	7	6	4	9	2	3	1
2	9	1	3	5	7	6	8	4
6	3	4	1	2	8	9	5	7

Solution # 676

4	9	3	5	1	8	6	7	2
8	5	7	2	6	3	1	9	4
6	2	1	7	9	4	8	5	3
3	6	5	8	2	7	4	1	9
7	1	8	4	5	9	3	2	6
9	4	2	6	3	1	7	8	5
1	3	4	9	7	5	2	6	8
2	8	9	1	4	6	5	3	7
5	7	6	3	8	2	9	4	1

Solution # 677

4	8	9	5	7	1	2	6	3
7	3	2	4	6	8	5	9	1
5	6	1	2	3	9	7	8	4
2	7	4	1	5	6	8	3	9
3	9	6	8	4	2	1	5	7
8	1	5	3	9	7	4	2	6
6	4	3	7	2	5	9	1	8
9	5	8	6	1	4	3	7	2
1	2	7	9	8	3	6	4	5

Solution # 678

7	1	5	6	9	4	2	3	8
2	9	4	3	1	8	7	6	5
3	8	6	2	5	7	4	9	1
4	5	3	1	2	6	8	7	9
8	2	7	5	4	9	3	1	6
1	6	9	7	8	3	5	4	2
9	7	8	4	6	5	1	2	3
5	4	2	9	3	1	6	8	7
6	3	1	8	7	2	9	5	4

Solution # 679

6	7	3	5	2	8	1	4	9
2	9	5	1	4	7	8	6	3
4	1	8	3	6	9	5	7	2
9	2	7	6	1	4	3	5	8
3	6	1	8	5	2	4	9	7
5	8	4	9	7	3	2	1	6
8	3	6	4	9	5	7	2	1
7	4	9	2	3	1	6	8	5
1	5	2	7	8	6	9	3	4

Solution # 680

7	9	8	4	5	2	6	3	1
4	5	1	3	6	7	8	9	2
2	3	6	9	8	1	5	4	7
5	4	3	7	1	9	2	8	6
1	2	7	6	4	8	3	5	9
8	6	9	2	3	5	7	1	4
6	1	4	8	2	3	9	7	5
3	7	2	5	9	4	1	6	8
9	8	5	1	7	6	4	2	3

Solution # 681

7	1	6	3	4	2	9	5	8
5	9	4	6	1	8	3	7	2
2	3	8	7	5	9	1	4	6
1	2	3	9	7	4	8	6	5
4	8	7	5	3	6	2	9	1
6	5	9	2	8	1	7	3	4
9	6	1	4	2	7	5	8	3
3	4	2	8	9	5	6	1	7
8	7	5	1	6	3	4	2	9

Solution # 682

4	3	8	7	2	9	1	5	6
2	6	1	3	4	5	8	7	9
9	7	5	6	8	1	3	4	2
5	2	9	8	6	4	7	3	1
1	8	7	2	5	3	9	6	4
3	4	6	1	9	7	5	2	8
6	5	2	9	3	8	4	1	7
7	9	3	4	1	6	2	8	5
8	1	4	5	7	2	6	9	3

Solution # 683

8	3	7	6	4	1	5	9	2
5	2	1	7	3	9	4	6	8
9	6	4	8	5	2	7	1	3
3	7	8	5	6	4	9	2	1
2	9	6	1	7	3	8	4	5
1	4	5	9	2	8	3	7	6
6	5	3	2	9	7	1	8	4
4	8	9	3	1	6	2	5	7
7	1	2	4	8	5	6	3	9

Solution # 684

6	3	4	1	8	9	2	7	5
9	5	1	7	2	4	3	6	8
2	8	7	3	5	6	1	9	4
5	7	3	4	6	8	9	1	2
1	4	9	2	7	3	5	8	6
8	6	2	9	1	5	4	3	7
7	9	8	5	3	2	6	4	1
3	1	5	6	4	7	8	2	9
4	2	6	8	9	1	7	5	3

Solution # 685

4	8	6	5	9	2	3	1	7
9	5	3	7	1	8	6	2	4
1	7	2	4	3	6	9	8	5
8	9	5	1	2	7	4	6	3
2	4	1	9	6	3	5	7	8
3	6	7	8	4	5	2	9	1
5	1	4	6	7	9	8	3	2
6	3	8	2	5	1	7	4	9
7	2	9	3	8	4	1	5	6

Solution # 686

8	4	1	9	2	7	6	3	5
2	3	6	8	1	5	4	7	9
5	9	7	6	3	4	2	1	8
9	5	2	7	8	3	1	6	4
1	8	3	4	6	9	7	5	2
7	6	4	2	5	1	8	9	3
4	1	5	3	7	8	9	2	6
6	7	8	5	9	2	3	4	1
3	2	9	1	4	6	5	8	7

Solution # 687

5	7	6	1	2	8	3	4	9
3	1	9	4	5	6	2	8	7
2	4	8	7	9	3	1	5	6
1	3	7	5	4	2	6	9	8
6	9	5	8	3	1	4	7	2
8	2	4	9	6	7	5	3	1
7	8	3	6	1	4	9	2	5
4	5	1	2	7	9	8	6	3
9	6	2	3	8	5	7	1	4

Solution # 688

9	5	4	7	3	8	6	1	2
3	2	8	6	9	1	4	7	5
6	7	1	5	4	2	9	8	3
4	1	5	9	7	3	8	2	6
2	9	3	1	8	6	5	4	7
8	6	7	2	5	4	1	3	9
1	3	2	4	6	9	7	5	8
5	4	9	8	2	7	3	6	1
7	8	6	3	1	5	2	9	4

Solution # 689

5	4	8	7	9	6	3	1	2
1	9	2	4	8	3	6	7	5
7	6	3	2	5	1	4	8	9
4	5	1	8	2	7	9	6	3
6	3	7	1	4	9	2	5	8
8	2	9	3	6	5	1	4	7
3	1	4	9	7	8	5	2	6
9	7	6	5	1	2	8	3	4
2	8	5	6	3	4	7	9	1

Solution # 690

9	6	5	1	3	4	2	7	8
4	2	3	8	7	5	1	9	6
1	7	8	2	9	6	4	5	3
7	4	6	5	8	9	3	2	1
8	5	2	4	1	3	9	6	7
3	9	1	6	2	7	8	4	5
6	8	4	3	5	2	7	1	9
5	1	7	9	4	8	6	3	2
2	3	9	7	6	1	5	8	4

Solution # 691

8	5	2	7	6	4	3	1	9
4	7	9	3	5	1	8	6	2
6	3	1	2	9	8	5	7	4
3	4	7	9	2	5	6	8	1
2	8	5	6	1	7	4	9	3
9	1	6	8	4	3	2	5	7
1	2	4	5	8	9	7	3	6
5	9	3	4	7	6	1	2	8
7	6	8	1	3	2	9	4	5

Solution # 692

9	3	6	5	4	7	2	1	8
4	5	7	8	2	1	6	9	3
1	2	8	9	3	6	4	7	5
5	4	2	1	8	9	3	6	7
3	8	9	6	7	2	1	5	4
6	7	1	3	5	4	9	8	2
7	9	3	2	6	5	8	4	1
8	6	4	7	1	3	5	2	9
2	1	5	4	9	8	7	3	6

Solution # 693

2	4	8	7	1	5	9	6	3
9	6	5	8	4	3	7	1	2
7	3	1	9	6	2	8	5	4
3	8	4	2	7	6	1	9	5
5	2	7	3	9	1	6	4	8
6	1	9	5	8	4	2	3	7
4	9	2	1	3	7	5	8	6
1	5	6	4	2	8	3	7	9
8	7	3	6	5	9	4	2	1

Solution # 694

9	7	1	5	3	2	6	8	4
8	3	4	9	7	6	5	2	1
5	6	2	4	1	8	3	9	7
3	4	7	2	8	1	9	5	6
2	9	5	7	6	4	1	3	8
6	1	8	3	5	9	4	7	2
4	8	3	1	9	7	2	6	5
1	5	6	8	2	3	7	4	9
7	2	9	6	4	5	8	1	3

Solution # 695

8	4	2	1	6	3	5	7	9
6	9	7	8	2	5	1	4	3
5	1	3	4	9	7	6	2	8
9	5	4	2	8	6	3	1	7
1	7	8	3	4	9	2	6	5
2	3	6	5	7	1	9	8	4
3	2	5	7	1	8	4	9	6
4	8	9	6	5	2	7	3	1
7	6	1	9	3	4	8	5	2

Solution # 696

8	5	7	6	3	4	9	1	2
6	9	2	5	1	8	4	7	3
1	3	4	9	2	7	6	8	5
5	6	9	3	4	1	8	2	7
3	2	8	7	9	6	1	5	4
4	7	1	2	8	5	3	9	6
9	8	3	4	5	2	7	6	1
2	1	6	8	7	3	5	4	9
7	4	5	1	6	9	2	3	8

Solution # 697

1	8	3	5	4	7	2	9	6
6	9	7	2	3	8	5	1	4
2	4	5	1	6	9	3	8	7
8	6	1	3	7	2	9	4	5
7	2	4	9	1	5	8	6	3
5	3	9	4	8	6	7	2	1
4	5	8	6	2	3	1	7	9
3	1	2	7	9	4	6	5	8
9	7	6	8	5	1	4	3	2

Solution # 698

8	3	5	9	1	7	2	4	6
2	9	1	4	6	3	7	8	5
6	4	7	2	5	8	3	1	9
3	5	4	1	2	9	6	7	8
9	7	6	5	8	4	1	3	2
1	2	8	7	3	6	5	9	4
7	8	3	6	9	5	4	2	1
5	1	9	3	4	2	8	6	7
4	6	2	8	7	1	9	5	3

Solution # 699

4	3	7	8	2	6	9	1	5
2	6	8	9	5	1	3	7	4
5	1	9	4	3	7	2	6	8
9	2	4	1	6	3	5	8	7
3	8	6	5	7	9	1	4	2
7	5	1	2	8	4	6	3	9
6	7	2	3	4	5	8	9	1
1	4	5	6	9	8	7	2	3
8	9	3	7	1	2	4	5	6

Solution # 700

2	8	7	3	1	6	4	5	9
3	9	1	4	5	2	7	6	8
5	6	4	9	8	7	1	3	2
1	2	9	6	4	3	5	8	7
4	3	5	8	7	1	2	9	6
6	7	8	2	9	5	3	4	1
8	4	2	7	3	9	6	1	5
9	1	6	5	2	4	8	7	3
7	5	3	1	6	8	9	2	4

Solution # 701

3	2	9	8	7	6	1	4	5
8	7	1	4	2	5	6	3	9
5	6	4	9	1	3	8	2	7
2	1	3	6	5	7	4	9	8
7	4	8	2	3	9	5	1	6
6	9	5	1	8	4	2	7	3
1	5	2	7	9	8	3	6	4
4	3	7	5	6	2	9	8	1
9	8	6	3	4	1	7	5	2

Solution # 702

8	1	7	9	3	6	5	2	4
3	2	5	7	1	4	6	9	8
4	9	6	5	8	2	7	3	1
7	3	1	2	4	9	8	6	5
5	4	8	3	6	7	2	1	9
2	6	9	1	5	8	4	7	3
9	5	2	4	7	1	3	8	6
6	7	4	8	9	3	1	5	2
1	8	3	6	2	5	9	4	7

Solution # 703

9	6	3	5	2	1	8	4	7
7	4	2	6	3	8	5	1	9
1	8	5	4	7	9	6	2	3
6	5	1	9	4	2	3	7	8
8	3	7	1	6	5	4	9	2
2	9	4	7	8	3	1	6	5
5	2	6	8	1	7	9	3	4
4	7	8	3	9	6	2	5	1
3	1	9	2	5	4	7	8	6

Solution # 704

5	4	9	6	7	3	2	8	1
3	7	8	1	4	2	9	5	6
6	2	1	8	5	9	3	4	7
4	9	6	7	3	8	5	1	2
2	1	5	4	9	6	7	3	8
8	3	7	5	2	1	4	6	9
1	5	4	2	6	7	8	9	3
7	8	3	9	1	5	6	2	4
9	6	2	3	8	4	1	7	5

Solution # 705

3	4	2	1	5	6	9	7	8
7	6	9	4	8	2	1	5	3
5	1	8	7	9	3	4	2	6
1	2	5	9	3	4	6	8	7
9	8	6	5	2	7	3	4	1
4	7	3	8	6	1	2	9	5
6	5	7	2	1	9	8	3	4
8	9	1	3	4	5	7	6	2
2	3	4	6	7	8	5	1	9

Solution # 706

4	5	8	7	1	3	9	2	6
7	1	2	5	9	6	3	8	4
3	9	6	2	8	4	5	7	1
2	6	3	1	5	8	4	9	7
8	7	5	6	4	9	2	1	3
9	4	1	3	7	2	8	6	5
1	8	9	4	6	5	7	3	2
5	2	7	9	3	1	6	4	8
6	3	4	8	2	7	1	5	9

Solution # 707

5	3	4	9	7	6	1	8	2
9	8	7	2	4	1	3	5	6
6	1	2	8	5	3	4	9	7
7	9	3	4	2	8	5	6	1
8	5	1	6	3	7	2	4	9
2	4	6	5	1	9	8	7	3
3	7	9	1	8	4	6	2	5
4	6	5	3	9	2	7	1	8
1	2	8	7	6	5	9	3	4

Solution # 708

5	4	2	3	9	7	6	8	1
9	8	1	6	4	5	7	2	3
7	6	3	8	2	1	5	9	4
3	5	6	2	7	9	4	1	8
2	1	7	4	6	8	9	3	5
4	9	8	5	1	3	2	6	7
8	2	9	7	3	4	1	5	6
6	3	4	1	5	2	8	7	9
1	7	5	9	8	6	3	4	2

Solution # 709

1	4	3	5	2	6	7	8	9
7	5	6	3	8	9	1	2	4
2	8	9	7	1	4	6	5	3
3	2	8	1	9	7	4	6	5
5	1	7	4	6	2	3	9	8
6	9	4	8	3	5	2	1	7
8	7	2	6	5	3	9	4	1
4	6	1	9	7	8	5	3	2
9	3	5	2	4	1	8	7	6

Solution # 710

4	2	7	3	5	8	1	9	6
6	5	1	2	4	9	8	7	3
3	8	9	7	6	1	4	5	2
1	4	8	6	3	5	7	2	9
9	6	5	1	7	2	3	8	4
2	7	3	9	8	4	5	6	1
8	3	4	5	2	6	9	1	7
5	9	6	4	1	7	2	3	8
7	1	2	8	9	3	6	4	5

Solution # 711

9	2	7	8	3	4	5	1	6
3	8	5	1	6	9	7	4	2
4	1	6	2	5	7	8	3	9
1	7	9	6	4	8	3	2	5
5	3	8	7	9	2	4	6	1
2	6	4	5	1	3	9	8	7
6	9	1	4	8	5	2	7	3
8	5	2	3	7	6	1	9	4
7	4	3	9	2	1	6	5	8

Solution # 712

7	4	9	3	1	6	8	5	2
2	3	8	4	7	5	1	9	6
5	6	1	2	9	8	3	4	7
1	2	3	9	6	4	5	7	8
4	8	7	5	3	2	9	6	1
9	5	6	1	8	7	2	3	4
8	1	4	7	5	9	6	2	3
6	9	2	8	4	3	7	1	5
3	7	5	6	2	1	4	8	9

Solution # 713

6	9	7	1	5	4	3	2	8
5	8	1	3	6	2	4	9	7
3	4	2	9	7	8	6	1	5
4	7	9	2	8	1	5	3	6
2	3	5	4	9	6	8	7	1
1	6	8	5	3	7	2	4	9
8	1	3	7	4	5	9	6	2
7	5	4	6	2	9	1	8	3
9	2	6	8	1	3	7	5	4

Solution # 714

1	8	5	7	3	9	4	6	2
4	6	9	8	2	1	3	5	7
2	3	7	5	6	4	9	8	1
3	9	4	2	8	7	6	1	5
8	5	1	6	9	3	2	7	4
6	7	2	4	1	5	8	3	9
9	4	8	1	5	6	7	2	3
7	1	6	3	4	2	5	9	8
5	2	3	9	7	8	1	4	6

Solution # 715

3	4	1	8	9	2	7	6	5
8	2	7	5	1	6	4	9	3
5	9	6	3	4	7	1	8	2
6	8	5	7	3	9	2	4	1
4	3	9	1	2	8	6	5	7
1	7	2	4	6	5	9	3	8
9	5	3	6	7	1	8	2	4
2	1	8	9	5	4	3	7	6
7	6	4	2	8	3	5	1	9

Solution # 716

6	5	7	3	1	9	8	2	4
3	4	8	6	7	2	9	5	1
9	2	1	5	4	8	6	7	3
5	9	2	7	6	4	1	3	8
7	1	3	8	9	5	2	4	6
4	8	6	2	3	1	7	9	5
2	7	5	1	8	3	4	6	9
8	6	9	4	5	7	3	1	2
1	3	4	9	2	6	5	8	7

Solution # 717

2	9	6	8	1	3	7	4	5
4	8	1	5	9	7	3	6	2
3	7	5	6	4	2	9	1	8
1	3	8	7	6	9	2	5	4
6	2	7	4	5	8	1	3	9
5	4	9	2	3	1	8	7	6
7	6	3	9	8	5	4	2	1
8	1	4	3	2	6	5	9	7
9	5	2	1	7	4	6	8	3

Solution # 718

4	8	1	6	9	3	7	5	2
2	9	7	4	5	1	6	8	3
5	6	3	2	7	8	4	1	9
8	7	5	1	3	4	9	2	6
1	4	2	9	6	5	8	3	7
9	3	6	7	8	2	1	4	5
3	5	9	8	4	7	2	6	1
6	2	4	5	1	9	3	7	8
7	1	8	3	2	6	5	9	4

Solution # 719

6	4	3	1	9	5	7	8	2
9	5	7	3	2	8	6	4	1
2	8	1	4	7	6	9	5	3
7	6	5	9	4	2	1	3	8
8	3	9	5	1	7	4	2	6
4	1	2	6	8	3	5	7	9
1	9	8	2	5	4	3	6	7
3	7	4	8	6	1	2	9	5
5	2	6	7	3	9	8	1	4

Solution # 720

9	2	6	3	4	7	1	5	8
5	4	8	1	2	6	9	7	3
7	3	1	5	8	9	6	4	2
4	5	7	2	9	1	8	3	6
2	8	3	6	7	5	4	9	1
1	6	9	8	3	4	7	2	5
6	7	2	9	1	3	5	8	4
3	9	5	4	6	8	2	1	7
8	1	4	7	5	2	3	6	9

Solution # 721

7	3	2	9	8	6	4	1	5
4	1	8	5	7	3	2	9	6
9	6	5	1	4	2	8	7	3
3	5	9	8	6	1	7	2	4
2	4	1	3	5	7	6	8	9
6	8	7	2	9	4	5	3	1
8	9	3	4	2	5	1	6	7
5	2	6	7	1	9	3	4	8
1	7	4	6	3	8	9	5	2

Solution # 722

3	7	5	8	2	1	6	4	9
1	9	2	4	6	5	3	8	7
8	4	6	3	9	7	2	5	1
7	2	4	6	1	3	5	9	8
9	5	3	2	4	8	7	1	6
6	1	8	5	7	9	4	3	2
4	3	9	7	8	6	1	2	5
5	6	1	9	3	2	8	7	4
2	8	7	1	5	4	9	6	3

Solution # 723

8	5	3	1	6	2	7	9	4
2	6	1	7	4	9	3	8	5
9	7	4	3	5	8	6	1	2
1	9	8	4	7	6	2	5	3
6	3	7	9	2	5	1	4	8
4	2	5	8	1	3	9	6	7
3	1	6	5	8	7	4	2	9
7	8	2	6	9	4	5	3	1
5	4	9	2	3	1	8	7	6

Solution # 724

2	7	4	6	3	1	8	9	5
3	5	1	9	8	4	2	6	7
8	9	6	5	7	2	1	4	3
5	1	9	8	2	6	3	7	4
4	8	7	3	1	5	9	2	6
6	3	2	7	4	9	5	1	8
9	2	8	4	5	7	6	3	1
1	4	5	2	6	3	7	8	9
7	6	3	1	9	8	4	5	2

Solution # 725

4	3	5	2	1	7	8	9	6
2	6	8	9	3	5	4	7	1
9	7	1	8	4	6	2	3	5
3	5	4	7	8	2	6	1	9
6	8	2	3	9	1	7	5	4
1	9	7	5	6	4	3	8	2
7	1	9	4	2	3	5	6	8
8	2	3	6	5	9	1	4	7
5	4	6	1	7	8	9	2	3

Solution # 726

4	2	6	3	9	7	8	5	1
9	1	5	8	6	2	4	3	7
3	8	7	1	5	4	9	2	6
6	4	3	7	1	5	2	8	9
5	9	1	2	4	8	6	7	3
8	7	2	6	3	9	1	4	5
1	3	8	5	2	6	7	9	4
7	6	4	9	8	3	5	1	2
2	5	9	4	7	1	3	6	8

Solution # 727

4	5	2	8	9	6	7	1	3
6	9	3	4	1	7	5	8	2
1	7	8	3	5	2	4	6	9
8	4	1	9	2	3	6	7	5
9	6	5	7	8	4	3	2	1
3	2	7	5	6	1	8	9	4
7	3	6	1	4	9	2	5	8
5	1	4	2	7	8	9	3	6
2	8	9	6	3	5	1	4	7

Solution # 728

4	3	9	6	8	1	7	2	5
5	2	6	4	7	3	9	1	8
7	8	1	5	2	9	6	4	3
3	1	8	9	6	7	4	5	2
9	6	5	1	4	2	3	8	7
2	4	7	8	3	5	1	9	6
8	5	4	7	9	6	2	3	1
1	7	3	2	5	4	8	6	9
6	9	2	3	1	8	5	7	4

Solution # 729

2	4	5	3	7	9	1	8	6
9	6	8	1	4	2	5	3	7
7	1	3	8	5	6	2	9	4
6	9	4	2	1	3	7	5	8
3	7	2	6	8	5	9	4	1
5	8	1	7	9	4	3	6	2
8	3	6	5	2	1	4	7	9
1	5	9	4	6	7	8	2	3
4	2	7	9	3	8	6	1	5

Solution # 730

4	3	7	1	2	5	9	8	6
5	1	9	4	8	6	7	3	2
8	2	6	7	9	3	4	1	5
7	8	4	9	5	1	2	6	3
9	6	1	2	3	8	5	7	4
2	5	3	6	4	7	8	9	1
3	4	5	8	6	9	1	2	7
6	7	8	5	1	2	3	4	9
1	9	2	3	7	4	6	5	8

Solution # 731

4	5	9	8	1	3	6	2	7
1	6	7	9	4	2	8	3	5
3	2	8	5	6	7	4	1	9
9	7	1	3	5	8	2	6	4
2	8	4	1	9	6	5	7	3
5	3	6	2	7	4	1	9	8
6	1	3	7	8	5	9	4	2
8	9	2	4	3	1	7	5	6
7	4	5	6	2	9	3	8	1

Solution # 732

3	9	7	4	2	5	1	8	6
5	8	6	3	1	7	2	4	9
2	4	1	8	6	9	7	3	5
8	5	4	2	7	3	9	6	1
9	6	2	1	5	8	4	7	3
7	1	3	6	9	4	8	5	2
1	3	8	5	4	2	6	9	7
6	7	5	9	8	1	3	2	4
4	2	9	7	3	6	5	1	8

Solution # 733

2	9	8	3	1	7	4	6	5
6	5	1	8	4	2	9	7	3
4	3	7	6	5	9	8	1	2
5	4	2	9	8	1	7	3	6
1	8	9	7	3	6	5	2	4
3	7	6	4	2	5	1	9	8
8	1	4	2	7	3	6	5	9
9	2	5	1	6	8	3	4	7
7	6	3	5	9	4	2	8	1

Solution # 734

4	5	1	2	7	3	8	6	9
7	8	3	9	6	1	2	4	5
9	6	2	4	8	5	7	3	1
3	2	9	5	1	4	6	8	7
1	7	8	3	9	6	4	5	2
6	4	5	8	2	7	9	1	3
5	3	7	6	4	2	1	9	8
2	9	6	1	5	8	3	7	4
8	1	4	7	3	9	5	2	6

Solution # 735

7	6	2	9	3	1	8	4	5
4	3	1	5	7	8	6	2	9
5	9	8	2	6	4	7	3	1
2	5	4	8	1	9	3	6	7
1	8	3	7	5	6	4	9	2
9	7	6	4	2	3	5	1	8
8	4	7	6	9	2	1	5	3
3	2	5	1	4	7	9	8	6
6	1	9	3	8	5	2	7	4

Solution # 736

7	8	1	6	5	9	3	2	4
2	5	9	1	4	3	7	8	6
6	3	4	7	8	2	1	9	5
1	7	6	3	9	4	8	5	2
4	2	8	5	7	1	6	3	9
5	9	3	2	6	8	4	1	7
3	6	2	9	1	7	5	4	8
8	1	7	4	2	5	9	6	3
9	4	5	8	3	6	2	7	1

Solution # 737

5	1	2	4	6	9	3	8	7
7	4	3	2	5	8	1	9	6
6	8	9	7	1	3	4	5	2
3	7	4	8	2	5	9	6	1
2	6	1	9	7	4	5	3	8
8	9	5	1	3	6	7	2	4
9	5	7	6	8	1	2	4	3
1	3	8	5	4	2	6	7	9
4	2	6	3	9	7	8	1	5

Solution # 738

8	6	1	5	9	7	3	2	4
4	3	7	2	6	1	9	5	8
2	9	5	8	3	4	1	7	6
6	8	4	7	1	9	5	3	2
5	7	9	3	8	2	6	4	1
3	1	2	4	5	6	7	8	9
1	2	8	9	7	3	4	6	5
7	4	6	1	2	5	8	9	3
9	5	3	6	4	8	2	1	7

Solution # 739

8	4	9	2	5	6	7	1	3
5	7	1	4	8	3	9	6	2
3	2	6	7	9	1	5	8	4
1	3	7	8	4	9	2	5	6
4	9	2	5	6	7	1	3	8
6	8	5	3	1	2	4	9	7
9	6	3	1	2	4	8	7	5
7	5	4	9	3	8	6	2	1
2	1	8	6	7	5	3	4	9

Solution # 740

4	6	3	2	7	9	8	5	1
5	8	2	6	1	3	4	7	9
7	9	1	8	5	4	2	3	6
1	7	5	3	9	2	6	8	4
6	2	4	1	8	7	3	9	5
8	3	9	5	4	6	7	1	2
9	1	6	7	2	8	5	4	3
3	5	8	4	6	1	9	2	7
2	4	7	9	3	5	1	6	8

Solution # 741

3	7	2	1	9	4	5	6	8
8	1	9	5	6	7	3	4	2
5	4	6	2	3	8	7	9	1
6	8	4	3	1	9	2	7	5
2	3	5	7	4	6	8	1	9
7	9	1	8	5	2	6	3	4
1	2	8	9	7	3	4	5	6
4	5	3	6	2	1	9	8	7
9	6	7	4	8	5	1	2	3

Solution # 742

4	7	3	5	9	6	1	8	2
8	5	6	1	3	2	4	9	7
2	1	9	4	8	7	6	3	5
9	2	4	8	5	1	3	7	6
5	6	8	3	7	9	2	1	4
7	3	1	6	2	4	8	5	9
6	9	2	7	1	8	5	4	3
1	4	5	9	6	3	7	2	8
3	8	7	2	4	5	9	6	1

Solution # 743

5	2	9	3	8	7	4	1	6
1	7	8	9	4	6	5	2	3
3	4	6	2	1	5	8	9	7
9	6	3	7	2	4	1	8	5
4	8	1	6	5	3	9	7	2
7	5	2	1	9	8	6	3	4
2	3	5	8	6	9	7	4	1
6	9	7	4	3	1	2	5	8
8	1	4	5	7	2	3	6	9

Solution # 744

4	8	1	5	6	9	3	7	2
5	2	6	7	4	3	1	9	8
9	7	3	8	1	2	4	5	6
7	9	2	1	3	5	6	8	4
3	5	4	6	9	8	2	1	7
1	6	8	4	2	7	5	3	9
8	3	5	2	7	4	9	6	1
2	1	7	9	5	6	8	4	3
6	4	9	3	8	1	7	2	5

Solution # 745

7	5	6	4	3	9	8	2	1
3	8	4	2	1	7	5	9	6
1	2	9	5	8	6	4	7	3
8	3	7	6	4	1	9	5	2
4	6	5	8	9	2	3	1	7
2	9	1	7	5	3	6	8	4
9	4	3	1	7	5	2	6	8
6	1	8	9	2	4	7	3	5
5	7	2	3	6	8	1	4	9

Solution # 746

1	3	8	9	7	6	5	4	2
7	5	6	8	4	2	9	3	1
9	2	4	3	5	1	7	6	8
2	8	5	6	1	4	3	7	9
3	7	1	5	9	8	4	2	6
4	6	9	2	3	7	1	8	5
5	1	2	7	8	3	6	9	4
8	4	3	1	6	9	2	5	7
6	9	7	4	2	5	8	1	3

Solution # 747

8	4	6	5	2	3	9	1	7
7	3	9	1	4	8	2	6	5
1	2	5	6	9	7	8	4	3
4	8	1	3	7	6	5	9	2
5	6	3	2	8	9	4	7	1
9	7	2	4	5	1	3	8	6
6	5	4	9	1	2	7	3	8
3	9	8	7	6	5	1	2	4
2	1	7	8	3	4	6	5	9

Solution # 748

9	2	6	8	4	7	3	5	1
1	4	3	9	6	5	2	8	7
7	5	8	2	3	1	4	6	9
2	7	4	5	9	8	1	3	6
5	6	1	7	2	3	9	4	8
3	8	9	4	1	6	5	7	2
6	3	2	1	7	4	8	9	5
8	9	7	3	5	2	6	1	4
4	1	5	6	8	9	7	2	3

Solution # 749

3	7	5	2	8	6	9	1	4
4	8	6	5	1	9	2	3	7
1	2	9	4	3	7	6	8	5
5	9	3	6	4	8	1	7	2
8	6	7	1	2	5	3	4	9
2	4	1	9	7	3	5	6	8
7	3	2	8	5	1	4	9	6
6	5	8	3	9	4	7	2	1
9	1	4	7	6	2	8	5	3

Solution # 750

2	5	4	7	6	3	1	9	8
7	3	9	4	1	8	2	6	5
1	8	6	9	5	2	3	7	4
5	2	1	6	3	7	8	4	9
3	9	8	5	4	1	7	2	6
6	4	7	8	2	9	5	1	3
4	6	3	2	7	5	9	8	1
8	1	2	3	9	4	6	5	7
9	7	5	1	8	6	4	3	2

Solution # 751

4	9	8	7	5	1	2	3	6
5	7	6	3	2	9	1	8	4
1	2	3	4	6	8	9	7	5
9	4	2	5	3	6	8	1	7
7	8	5	1	4	2	3	6	9
6	3	1	9	8	7	5	4	2
3	5	7	8	9	4	6	2	1
8	6	4	2	1	5	7	9	3
2	1	9	6	7	3	4	5	8

Solution # 752

7	2	4	5	1	8	6	3	9
5	6	1	3	9	2	4	7	8
9	8	3	4	7	6	5	1	2
2	5	9	8	3	4	7	6	1
8	3	6	1	2	7	9	4	5
1	4	7	9	6	5	2	8	3
4	1	2	7	5	3	8	9	6
3	7	5	6	8	9	1	2	4
6	9	8	2	4	1	3	5	7

Solution # 753

4	5	6	8	7	3	9	2	1
8	2	9	1	4	6	3	7	5
3	7	1	9	5	2	4	8	6
1	3	7	2	6	4	5	9	8
6	8	5	7	9	1	2	3	4
2	9	4	3	8	5	6	1	7
5	1	3	6	2	8	7	4	9
7	6	8	4	3	9	1	5	2
9	4	2	5	1	7	8	6	3

Solution # 754

5	1	9	6	8	7	3	4	2
8	3	6	5	2	4	1	7	9
4	7	2	3	1	9	5	6	8
2	4	8	7	5	1	6	9	3
7	9	5	8	3	6	2	1	4
3	6	1	9	4	2	7	8	5
1	8	3	4	7	5	9	2	6
6	5	7	2	9	8	4	3	1
9	2	4	1	6	3	8	5	7

Solution # 755

1	4	6	8	9	5	2	7	3
3	9	8	2	4	7	1	5	6
2	7	5	3	1	6	9	4	8
5	3	4	1	6	2	8	9	7
9	6	2	5	7	8	4	3	1
8	1	7	4	3	9	5	6	2
4	5	1	7	2	3	6	8	9
7	8	9	6	5	1	3	2	4
6	2	3	9	8	4	7	1	5

Solution # 756

4	2	8	6	9	7	3	1	5
5	9	6	4	1	3	7	2	8
7	1	3	5	8	2	9	4	6
8	3	7	1	6	4	5	9	2
1	6	9	2	7	5	4	8	3
2	4	5	8	3	9	1	6	7
9	8	2	7	5	1	6	3	4
3	5	4	9	2	6	8	7	1
6	7	1	3	4	8	2	5	9

Solution # 757

7	6	8	5	2	1	3	4	9
5	3	4	6	7	9	2	8	1
9	2	1	3	8	4	6	5	7
6	5	7	9	4	2	1	3	8
8	9	3	7	1	6	4	2	5
1	4	2	8	5	3	7	9	6
4	7	5	1	3	8	9	6	2
2	8	6	4	9	7	5	1	3
3	1	9	2	6	5	8	7	4

Solution # 758

6	9	3	7	4	2	8	5	1
7	1	2	3	5	8	6	4	9
4	5	8	6	9	1	3	7	2
3	4	7	8	1	9	2	6	5
2	6	1	4	7	5	9	8	3
5	8	9	2	6	3	7	1	4
1	7	4	9	3	6	5	2	8
9	2	5	1	8	7	4	3	6
8	3	6	5	2	4	1	9	7

Solution # 759

1	5	3	4	6	9	8	2	7
7	9	6	2	8	1	5	4	3
2	8	4	5	3	7	1	9	6
4	1	8	7	2	3	9	6	5
9	7	5	6	4	8	3	1	2
3	6	2	1	9	5	4	7	8
6	4	9	8	5	2	7	3	1
8	3	7	9	1	6	2	5	4
5	2	1	3	7	4	6	8	9

Solution # 760

9	3	7	5	1	2	4	8	6
2	6	1	4	8	3	9	5	7
5	4	8	7	6	9	2	1	3
7	2	9	3	5	1	6	4	8
3	5	6	9	4	8	1	7	2
8	1	4	6	2	7	5	3	9
1	9	3	2	7	4	8	6	5
4	7	5	8	9	6	3	2	1
6	8	2	1	3	5	7	9	4

Solution # 761

8	6	4	9	2	5	7	1	3
9	1	5	7	6	3	4	8	2
3	7	2	1	4	8	5	9	6
1	9	6	5	8	2	3	7	4
4	3	7	6	9	1	8	2	5
5	2	8	3	7	4	1	6	9
6	4	9	8	3	7	2	5	1
2	8	1	4	5	6	9	3	7
7	5	3	2	1	9	6	4	8

Solution # 762

7	9	4	3	2	6	8	5	1
5	8	2	7	1	4	3	9	6
3	6	1	5	8	9	2	7	4
4	3	7	2	5	8	6	1	9
1	5	9	6	4	3	7	2	8
6	2	8	1	9	7	4	3	5
9	7	5	8	6	2	1	4	3
2	4	6	9	3	1	5	8	7
8	1	3	4	7	5	9	6	2

Solution # 763

3	1	7	2	5	8	6	9	4
4	8	2	3	6	9	7	5	1
5	9	6	4	1	7	3	2	8
7	2	4	6	9	5	8	1	3
1	3	5	8	7	4	2	6	9
9	6	8	1	3	2	5	4	7
8	5	3	9	4	6	1	7	2
2	7	9	5	8	1	4	3	6
6	4	1	7	2	3	9	8	5

Solution # 764

1	7	9	3	4	2	5	6	8
5	2	4	6	8	7	3	1	9
8	6	3	1	5	9	7	4	2
2	3	5	4	9	8	6	7	1
4	9	1	7	3	6	2	8	5
6	8	7	5	2	1	9	3	4
3	4	6	9	1	5	8	2	7
9	1	2	8	7	3	4	5	6
7	5	8	2	6	4	1	9	3

Solution # 765

1	4	5	8	9	7	3	6	2
7	2	9	4	6	3	8	1	5
8	3	6	5	1	2	7	4	9
2	9	4	1	7	6	5	8	3
3	7	8	9	4	5	6	2	1
6	5	1	3	2	8	9	7	4
5	6	7	2	3	1	4	9	8
9	8	2	7	5	4	1	3	6
4	1	3	6	8	9	2	5	7

Solution # 766

8	7	5	4	2	1	9	3	6
2	6	4	3	9	7	5	1	8
1	3	9	6	5	8	2	4	7
7	1	2	9	3	6	4	8	5
4	9	8	2	7	5	1	6	3
3	5	6	1	8	4	7	2	9
9	8	1	7	6	2	3	5	4
5	4	3	8	1	9	6	7	2
6	2	7	5	4	3	8	9	1

Solution # 767

7	5	9	1	3	2	8	4	6
4	2	1	5	6	8	3	9	7
8	6	3	4	9	7	1	2	5
1	4	7	2	8	5	6	3	9
2	9	8	3	4	6	5	7	1
5	3	6	7	1	9	4	8	2
3	7	4	9	5	1	2	6	8
9	8	5	6	2	3	7	1	4
6	1	2	8	7	4	9	5	3

Solution # 768

7	8	3	4	2	1	5	9	6
4	2	5	6	9	3	7	8	1
9	6	1	7	5	8	2	4	3
1	4	6	9	3	7	8	2	5
2	5	7	8	6	4	3	1	9
8	3	9	5	1	2	6	7	4
6	7	8	3	4	9	1	5	2
3	1	4	2	7	5	9	6	8
5	9	2	1	8	6	4	3	7

Solution # 769

3	8	6	1	2	5	4	9	7
4	7	5	9	3	8	6	1	2
2	9	1	7	4	6	8	5	3
7	3	9	4	5	1	2	8	6
8	1	4	2	6	9	3	7	5
5	6	2	3	8	7	1	4	9
9	2	3	8	7	4	5	6	1
6	4	7	5	1	2	9	3	8
1	5	8	6	9	3	7	2	4

Solution # 770

5	1	2	9	8	6	7	4	3
3	7	9	4	1	5	8	2	6
6	4	8	3	7	2	9	5	1
8	6	5	1	9	3	2	7	4
1	2	7	6	4	8	5	3	9
4	9	3	2	5	7	1	6	8
9	8	6	5	2	4	3	1	7
7	5	4	8	3	1	6	9	2
2	3	1	7	6	9	4	8	5

Solution # 771

1	4	5	2	9	6	3	8	7
7	2	9	1	8	3	6	5	4
8	6	3	4	5	7	9	1	2
4	1	8	9	7	5	2	6	3
5	9	6	3	4	2	8	7	1
2	3	7	8	6	1	4	9	5
6	5	2	7	3	8	1	4	9
3	8	4	5	1	9	7	2	6
9	7	1	6	2	4	5	3	8

Solution # 772

4	3	7	2	6	5	9	1	8
5	9	2	7	1	8	4	6	3
1	8	6	3	4	9	5	7	2
8	1	3	5	2	7	6	4	9
9	7	5	4	3	6	8	2	1
2	6	4	9	8	1	3	5	7
7	2	9	6	5	3	1	8	4
3	5	1	8	7	4	2	9	6
6	4	8	1	9	2	7	3	5

Solution # 773

3	4	6	1	9	5	7	2	8
2	9	1	4	7	8	5	6	3
5	7	8	6	3	2	1	4	9
4	6	2	3	8	1	9	7	5
8	3	7	9	5	4	6	1	2
9	1	5	7	2	6	3	8	4
1	8	3	5	4	7	2	9	6
6	5	4	2	1	9	8	3	7
7	2	9	8	6	3	4	5	1

Solution # 774

9	2	4	3	1	7	8	5	6
6	7	5	2	4	8	9	3	1
3	1	8	6	9	5	7	2	4
7	3	2	4	5	1	6	8	9
8	5	6	7	3	9	4	1	2
4	9	1	8	2	6	5	7	3
5	4	3	9	7	2	1	6	8
2	6	7	1	8	4	3	9	5
1	8	9	5	6	3	2	4	7

Solution # 775

8	3	6	2	4	9	7	5	1
5	7	2	8	6	1	3	4	9
1	9	4	3	7	5	2	8	6
7	4	9	6	5	2	1	3	8
6	5	1	4	3	8	9	2	7
3	2	8	9	1	7	5	6	4
4	8	7	1	2	3	6	9	5
9	1	3	5	8	6	4	7	2
2	6	5	7	9	4	8	1	3

Solution # 776

9	7	5	8	3	6	4	2	1
6	3	4	9	1	2	7	8	5
2	1	8	4	5	7	6	9	3
5	8	6	3	2	4	9	1	7
3	4	9	7	8	1	2	5	6
7	2	1	5	6	9	8	3	4
8	6	3	2	4	5	1	7	9
1	5	7	6	9	8	3	4	2
4	9	2	1	7	3	5	6	8

Solution # 777

4	6	9	7	3	1	2	8	5
5	7	1	2	6	8	4	9	3
2	3	8	4	9	5	7	1	6
1	4	7	5	8	3	9	6	2
6	5	2	9	4	7	1	3	8
9	8	3	6	1	2	5	7	4
3	1	4	8	5	9	6	2	7
8	2	5	1	7	6	3	4	9
7	9	6	3	2	4	8	5	1

Solution # 778

2	4	9	5	7	8	3	6	1
1	7	5	6	4	3	2	8	9
3	8	6	9	2	1	4	5	7
7	6	2	3	8	9	5	1	4
9	3	1	4	5	6	7	2	8
8	5	4	7	1	2	9	3	6
4	1	7	2	6	5	8	9	3
5	9	8	1	3	4	6	7	2
6	2	3	8	9	7	1	4	5

Solution # 779

1	3	4	2	7	6	9	5	8
7	8	5	9	1	3	6	2	4
9	2	6	4	5	8	7	3	1
3	6	2	7	4	1	8	9	5
4	1	8	3	9	5	2	7	6
5	7	9	8	6	2	4	1	3
6	9	7	1	3	4	5	8	2
2	4	1	5	8	7	3	6	9
8	5	3	6	2	9	1	4	7

Solution # 780

5	4	9	7	3	6	1	2	8
7	6	1	8	2	9	4	3	5
8	3	2	5	4	1	7	6	9
4	5	8	1	9	3	6	7	2
6	1	7	2	8	5	3	9	4
9	2	3	4	6	7	8	5	1
2	7	4	3	5	8	9	1	6
3	8	6	9	1	2	5	4	7
1	9	5	6	7	4	2	8	3

Solution # 781

5	7	1	6	3	2	9	8	4
4	2	9	1	5	8	3	6	7
3	8	6	7	4	9	2	1	5
9	5	7	3	8	6	4	2	1
1	4	2	5	9	7	8	3	6
8	6	3	4	2	1	5	7	9
7	3	4	8	1	5	6	9	2
6	9	8	2	7	4	1	5	3
2	1	5	9	6	3	7	4	8

Solution # 782

7	9	3	5	1	2	6	4	8
2	1	6	4	7	8	5	9	3
4	8	5	6	3	9	2	7	1
9	7	4	8	5	1	3	2	6
8	5	2	9	6	3	4	1	7
6	3	1	7	2	4	9	8	5
1	2	8	3	9	5	7	6	4
3	6	9	1	4	7	8	5	2
5	4	7	2	8	6	1	3	9

Solution # 783

8	1	3	9	2	7	5	6	4
6	4	2	8	1	5	9	3	7
5	9	7	6	4	3	8	2	1
3	5	1	7	6	4	2	9	8
4	6	9	3	8	2	1	7	5
7	2	8	5	9	1	3	4	6
1	8	6	4	3	9	7	5	2
2	3	5	1	7	6	4	8	9
9	7	4	2	5	8	6	1	3

Solution # 784

9	7	3	5	6	4	2	1	8
5	4	8	1	7	2	9	3	6
2	1	6	8	9	3	4	7	5
8	2	5	3	1	9	7	6	4
3	6	1	7	4	5	8	9	2
4	9	7	2	8	6	3	5	1
1	5	4	9	2	7	6	8	3
6	3	9	4	5	8	1	2	7
7	8	2	6	3	1	5	4	9

Solution # 785

3	2	5	8	1	4	7	6	9
4	1	6	7	5	9	8	2	3
8	7	9	2	6	3	1	5	4
9	8	7	5	4	2	3	1	6
6	5	1	3	9	7	4	8	2
2	3	4	1	8	6	9	7	5
5	9	3	6	7	8	2	4	1
1	4	8	9	2	5	6	3	7
7	6	2	4	3	1	5	9	8

Solution # 786

5	3	8	9	7	6	1	2	4
2	9	4	1	5	8	7	6	3
7	1	6	3	2	4	9	8	5
3	7	2	6	4	1	5	9	8
4	5	9	2	8	7	3	1	6
8	6	1	5	3	9	4	7	2
9	4	5	8	1	2	6	3	7
1	8	3	7	6	5	2	4	9
6	2	7	4	9	3	8	5	1

Solution # 787

6	5	2	3	1	7	9	4	8
7	3	1	4	8	9	5	6	2
4	8	9	5	2	6	7	3	1
2	7	3	6	5	8	1	9	4
5	4	6	9	7	1	2	8	3
1	9	8	2	4	3	6	7	5
9	2	5	8	6	4	3	1	7
8	6	7	1	3	5	4	2	9
3	1	4	7	9	2	8	5	6

Solution # 788

3	5	6	4	7	9	1	2	8
9	1	8	3	2	6	7	5	4
7	4	2	1	8	5	6	3	9
6	7	4	5	3	1	8	9	2
2	9	1	6	4	8	3	7	5
8	3	5	2	9	7	4	1	6
1	2	3	9	6	4	5	8	7
4	8	9	7	5	3	2	6	1
5	6	7	8	1	2	9	4	3

Solution # 789

8	9	4	6	5	2	3	1	7
1	2	7	4	3	9	6	8	5
6	3	5	7	1	8	9	2	4
3	5	6	8	7	1	2	4	9
9	7	1	2	4	3	8	5	6
2	4	8	5	9	6	1	7	3
4	6	9	1	2	7	5	3	8
5	1	3	9	8	4	7	6	2
7	8	2	3	6	5	4	9	1

Solution # 790

4	1	9	6	3	8	7	2	5
5	8	3	7	1	2	6	4	9
2	6	7	9	4	5	8	1	3
1	5	6	2	8	4	9	3	7
3	9	4	1	5	7	2	8	6
7	2	8	3	9	6	4	5	1
6	3	1	4	2	9	5	7	8
9	4	5	8	7	3	1	6	2
8	7	2	5	6	1	3	9	4

Solution # 791

2	5	1	8	7	4	9	6	3
9	4	6	3	2	5	7	1	8
3	7	8	1	9	6	5	4	2
7	3	2	6	5	8	4	9	1
8	1	5	9	4	7	3	2	6
4	6	9	2	1	3	8	7	5
5	2	3	7	6	9	1	8	4
6	9	4	5	8	1	2	3	7
1	8	7	4	3	2	6	5	9

Solution # 792

8	2	7	9	6	4	5	1	3
9	6	1	3	7	5	4	2	8
3	4	5	2	8	1	9	7	6
5	7	8	4	3	6	1	9	2
6	1	9	7	5	2	8	3	4
2	3	4	8	1	9	7	6	5
1	9	2	5	4	3	6	8	7
4	8	3	6	9	7	2	5	1
7	5	6	1	2	8	3	4	9

Solution # 793

8	6	2	3	7	4	5	1	9
1	5	3	9	8	2	7	6	4
4	9	7	5	1	6	2	8	3
3	4	6	8	9	5	1	2	7
7	2	9	6	3	1	8	4	5
5	8	1	4	2	7	9	3	6
2	7	5	1	6	3	4	9	8
9	3	4	2	5	8	6	7	1
6	1	8	7	4	9	3	5	2

Solution # 794

7	3	4	8	1	2	6	9	5
6	1	2	5	9	4	3	7	8
5	8	9	7	3	6	4	1	2
4	7	3	9	2	1	5	8	6
9	2	5	6	7	8	1	3	4
8	6	1	3	4	5	7	2	9
1	9	6	4	8	3	2	5	7
2	4	7	1	5	9	8	6	3
3	5	8	2	6	7	9	4	1

Solution # 795

4	5	1	3	8	2	6	7	9
9	2	6	7	4	5	8	1	3
7	8	3	6	9	1	4	2	5
5	9	7	8	2	3	1	4	6
1	4	8	5	7	6	9	3	2
3	6	2	9	1	4	5	8	7
8	3	9	1	6	7	2	5	4
2	1	5	4	3	9	7	6	8
6	7	4	2	5	8	3	9	1

Solution # 796

5	7	3	4	8	9	2	1	6
4	9	6	2	1	5	3	8	7
8	2	1	3	6	7	5	9	4
2	1	8	5	7	3	6	4	9
3	6	4	8	9	2	7	5	1
7	5	9	1	4	6	8	2	3
9	8	2	7	3	4	1	6	5
6	3	5	9	2	1	4	7	8
1	4	7	6	5	8	9	3	2

Solution # 797

6	9	5	3	2	8	7	4	1
4	2	3	6	1	7	9	8	5
1	8	7	5	9	4	6	2	3
3	6	8	9	7	1	2	5	4
7	4	9	2	3	5	1	6	8
5	1	2	8	4	6	3	7	9
8	5	1	7	6	9	4	3	2
9	3	6	4	5	2	8	1	7
2	7	4	1	8	3	5	9	6

Solution # 798

9	4	5	8	2	1	7	6	3
6	3	1	7	5	4	8	2	9
8	2	7	3	9	6	5	4	1
7	5	9	2	6	3	1	8	4
2	1	3	4	7	8	6	9	5
4	6	8	9	1	5	3	7	2
5	7	2	6	3	9	4	1	8
1	8	6	5	4	2	9	3	7
3	9	4	1	8	7	2	5	6

Solution # 799

5	8	4	2	3	6	1	9	7
2	6	3	1	9	7	8	5	4
7	1	9	8	4	5	2	3	6
8	9	2	4	7	3	5	6	1
6	5	1	9	8	2	4	7	3
4	3	7	6	5	1	9	2	8
9	7	5	3	1	4	6	8	2
3	4	6	5	2	8	7	1	9
1	2	8	7	6	9	3	4	5

Solution # 800

8	1	6	3	7	9	4	2	5
2	7	5	8	4	6	9	3	1
4	3	9	2	5	1	8	7	6
7	5	4	6	2	3	1	8	9
1	9	2	4	8	7	5	6	3
3	6	8	1	9	5	7	4	2
5	2	7	9	6	8	3	1	4
6	8	3	5	1	4	2	9	7
9	4	1	7	3	2	6	5	8

Solution # 801

8	4	1	7	2	3	9	5	6
5	6	3	4	9	8	7	1	2
9	2	7	1	6	5	3	4	8
7	5	8	9	4	1	2	6	3
1	3	4	2	8	6	5	9	7
2	9	6	3	5	7	4	8	1
4	7	9	6	1	2	8	3	5
3	1	5	8	7	9	6	2	4
6	8	2	5	3	4	1	7	9

Solution # 802

1	3	2	8	7	6	4	5	9
6	7	9	2	4	5	1	8	3
5	4	8	3	1	9	2	6	7
8	1	5	7	6	3	9	4	2
3	9	6	1	2	4	5	7	8
4	2	7	9	5	8	6	3	1
7	6	3	5	9	1	8	2	4
9	8	4	6	3	2	7	1	5
2	5	1	4	8	7	3	9	6

Solution # 803

2	3	7	6	4	9	5	8	1
1	5	9	2	8	7	6	4	3
8	6	4	3	1	5	7	2	9
7	1	2	9	6	8	3	5	4
6	4	8	5	7	3	1	9	2
5	9	3	4	2	1	8	7	6
4	2	1	8	5	6	9	3	7
3	7	5	1	9	4	2	6	8
9	8	6	7	3	2	4	1	5

Solution # 804

1	8	3	6	4	9	5	7	2
6	7	5	8	2	3	1	9	4
2	9	4	5	7	1	8	6	3
8	5	1	9	3	7	2	4	6
9	4	6	1	8	2	7	3	5
3	2	7	4	6	5	9	8	1
4	6	2	7	5	8	3	1	9
7	3	9	2	1	6	4	5	8
5	1	8	3	9	4	6	2	7

Solution # 805

6	4	7	3	1	8	5	9	2
2	3	1	7	5	9	8	6	4
5	8	9	6	2	4	7	1	3
7	9	5	4	3	6	1	2	8
4	1	2	9	8	5	6	3	7
8	6	3	2	7	1	9	4	5
9	5	6	8	4	2	3	7	1
1	7	4	5	9	3	2	8	6
3	2	8	1	6	7	4	5	9

Solution # 806

5	7	8	6	2	1	3	4	9
9	2	1	3	5	4	6	7	8
4	6	3	7	8	9	2	1	5
6	4	2	9	3	7	5	8	1
1	9	5	4	6	8	7	2	3
8	3	7	5	1	2	4	9	6
2	8	6	1	4	5	9	3	7
3	1	9	2	7	6	8	5	4
7	5	4	8	9	3	1	6	2

Solution # 807

4	9	7	8	2	1	3	6	5
2	1	6	4	3	5	7	8	9
3	5	8	6	7	9	2	4	1
8	2	5	3	4	6	1	9	7
7	6	9	1	5	8	4	2	3
1	3	4	7	9	2	6	5	8
5	8	1	2	6	3	9	7	4
6	7	3	9	8	4	5	1	2
9	4	2	5	1	7	8	3	6

Solution # 808

9	2	5	1	3	6	4	7	8
3	4	7	8	9	5	2	1	6
6	1	8	4	7	2	5	9	3
1	9	4	6	2	3	8	5	7
5	6	3	7	4	8	1	2	9
7	8	2	9	5	1	3	6	4
2	3	9	5	6	4	7	8	1
4	7	1	2	8	9	6	3	5
8	5	6	3	1	7	9	4	2

Solution # 809

6	2	3	5	1	4	9	7	8
5	4	9	7	8	3	1	6	2
8	1	7	2	6	9	3	5	4
1	3	8	9	5	6	4	2	7
4	7	5	8	2	1	6	9	3
2	9	6	3	4	7	5	8	1
3	6	2	4	9	8	7	1	5
9	5	4	1	7	2	8	3	6
7	8	1	6	3	5	2	4	9

Solution # 810

7	5	2	1	3	6	9	8	4
8	3	1	4	5	9	2	7	6
6	4	9	2	8	7	5	1	3
4	9	5	7	2	8	3	6	1
3	2	8	6	1	4	7	5	9
1	7	6	3	9	5	8	4	2
5	8	4	9	6	2	1	3	7
9	6	3	8	7	1	4	2	5
2	1	7	5	4	3	6	9	8

Solution # 811

2	4	6	9	3	5	7	8	1
9	3	1	4	8	7	6	5	2
8	7	5	1	2	6	4	9	3
5	6	7	2	9	3	8	1	4
4	9	8	7	6	1	2	3	5
3	1	2	5	4	8	9	6	7
7	8	3	6	5	2	1	4	9
6	2	4	3	1	9	5	7	8
1	5	9	8	7	4	3	2	6

Solution # 812

2	9	4	3	1	7	5	6	8
8	7	6	5	4	2	1	3	9
3	5	1	9	8	6	4	7	2
4	6	2	8	7	9	3	1	5
9	1	3	6	5	4	8	2	7
7	8	5	1	2	3	6	9	4
1	3	8	2	9	5	7	4	6
6	4	9	7	3	8	2	5	1
5	2	7	4	6	1	9	8	3

Solution # 813

4	5	9	8	6	2	3	7	1
2	8	7	1	4	3	6	9	5
6	3	1	5	7	9	2	4	8
5	9	3	2	1	8	7	6	4
7	1	2	4	3	6	5	8	9
8	4	6	7	9	5	1	2	3
3	7	5	6	8	4	9	1	2
1	2	4	9	5	7	8	3	6
9	6	8	3	2	1	4	5	7

Solution # 814

7	1	5	3	2	4	8	6	9
8	3	4	6	9	1	2	7	5
2	6	9	8	7	5	4	3	1
6	7	3	5	4	8	1	9	2
4	2	1	7	3	9	5	8	6
5	9	8	2	1	6	3	4	7
3	5	7	4	6	2	9	1	8
1	4	2	9	8	7	6	5	3
9	8	6	1	5	3	7	2	4

Solution # 815

8	1	7	4	2	5	6	3	9
5	6	4	3	9	1	7	2	8
2	9	3	7	6	8	1	4	5
4	8	2	5	1	7	3	9	6
6	3	5	9	8	2	4	1	7
9	7	1	6	3	4	5	8	2
3	4	9	2	7	6	8	5	1
1	5	6	8	4	9	2	7	3
7	2	8	1	5	3	9	6	4

Solution # 816

8	7	9	5	4	2	3	6	1
4	6	3	7	1	9	2	8	5
2	5	1	3	6	8	9	7	4
1	4	5	6	2	7	8	9	3
9	2	8	1	3	5	7	4	6
7	3	6	9	8	4	1	5	2
5	8	2	4	9	1	6	3	7
3	9	4	2	7	6	5	1	8
6	1	7	8	5	3	4	2	9

Solution # 817

8	9	7	1	2	5	3	6	4
2	4	5	7	6	3	9	1	8
6	1	3	9	4	8	5	2	7
4	2	1	5	3	9	7	8	6
3	8	6	4	7	1	2	5	9
7	5	9	2	8	6	1	4	3
9	6	2	3	5	4	8	7	1
5	3	4	8	1	7	6	9	2
1	7	8	6	9	2	4	3	5

Solution # 818

9	6	3	5	2	4	7	1	8
5	7	4	1	3	8	6	9	2
8	2	1	7	9	6	4	5	3
4	9	2	3	6	1	5	8	7
6	3	5	8	7	2	9	4	1
7	1	8	9	4	5	3	2	6
2	5	6	4	8	7	1	3	9
3	4	7	2	1	9	8	6	5
1	8	9	6	5	3	2	7	4

Solution # 819

4	2	7	1	9	8	5	6	3
5	3	9	2	7	6	4	1	8
8	6	1	3	5	4	2	9	7
9	8	4	7	1	2	3	5	6
2	5	6	9	4	3	8	7	1
1	7	3	6	8	5	9	2	4
6	4	8	5	2	7	1	3	9
7	1	5	8	3	9	6	4	2
3	9	2	4	6	1	7	8	5

Solution # 820

7	8	4	2	9	5	1	3	6
5	3	9	8	6	1	4	2	7
6	2	1	3	7	4	5	9	8
3	7	2	1	5	9	8	6	4
4	6	5	7	8	3	9	1	2
9	1	8	4	2	6	7	5	3
2	4	3	9	1	8	6	7	5
1	5	7	6	4	2	3	8	9
8	9	6	5	3	7	2	4	1

Solution # 821

8	6	3	1	2	5	9	4	7
4	1	9	3	6	7	5	8	2
5	2	7	9	4	8	1	6	3
7	5	6	8	3	2	4	9	1
2	9	4	5	1	6	3	7	8
1	3	8	7	9	4	2	5	6
3	7	1	4	8	9	6	2	5
6	4	5	2	7	1	8	3	9
9	8	2	6	5	3	7	1	4

Solution # 822

3	7	4	6	9	5	1	8	2
1	6	2	3	4	8	9	5	7
8	9	5	1	7	2	6	4	3
4	8	6	5	2	3	7	1	9
7	1	3	9	6	4	5	2	8
5	2	9	7	8	1	3	6	4
9	5	8	4	3	6	2	7	1
6	4	7	2	1	9	8	3	5
2	3	1	8	5	7	4	9	6

Solution # 823

1	3	4	2	9	8	7	5	6
9	6	7	4	3	5	1	2	8
5	2	8	1	7	6	9	4	3
2	5	3	9	6	7	4	8	1
4	1	6	3	8	2	5	9	7
8	7	9	5	1	4	3	6	2
3	9	2	8	5	1	6	7	4
7	4	1	6	2	9	8	3	5
6	8	5	7	4	3	2	1	9

Solution # 824

1	5	3	8	6	4	9	2	7
8	7	4	5	9	2	6	3	1
2	6	9	3	7	1	4	5	8
5	4	8	1	3	7	2	6	9
3	9	2	6	8	5	7	1	4
7	1	6	4	2	9	5	8	3
6	8	7	9	5	3	1	4	2
9	3	1	2	4	6	8	7	5
4	2	5	7	1	8	3	9	6

Solution # 825

5	4	1	9	6	7	3	8	2
9	8	3	2	4	5	6	7	1
7	6	2	8	1	3	5	4	9
2	7	8	1	5	9	4	6	3
3	1	6	4	7	2	8	9	5
4	9	5	6	3	8	2	1	7
6	3	4	7	2	1	9	5	8
8	5	7	3	9	6	1	2	4
1	2	9	5	8	4	7	3	6

Solution # 826

7	3	8	5	9	1	4	6	2
9	5	6	3	4	2	7	1	8
2	4	1	8	6	7	3	5	9
3	2	7	1	8	4	6	9	5
6	1	4	2	5	9	8	7	3
5	8	9	7	3	6	1	2	4
1	9	5	4	7	8	2	3	6
4	6	2	9	1	3	5	8	7
8	7	3	6	2	5	9	4	1

Solution # 827

2	1	4	7	5	8	9	6	3
5	8	7	9	3	6	4	2	1
3	6	9	1	4	2	5	8	7
9	5	8	6	2	1	3	7	4
4	3	6	8	7	5	1	9	2
1	7	2	3	9	4	6	5	8
6	2	5	4	8	3	7	1	9
7	4	1	2	6	9	8	3	5
8	9	3	5	1	7	2	4	6

Solution # 828

4	3	8	2	7	5	6	9	1
6	1	5	3	9	4	8	2	7
7	2	9	8	6	1	3	4	5
2	4	7	5	8	6	1	3	9
5	9	3	7	1	2	4	8	6
1	8	6	4	3	9	7	5	2
9	7	4	1	5	3	2	6	8
8	5	2	6	4	7	9	1	3
3	6	1	9	2	8	5	7	4

Solution # 829

3	7	9	1	8	2	4	6	5
5	4	8	9	3	6	7	2	1
2	6	1	4	5	7	9	3	8
6	3	2	5	7	1	8	9	4
1	9	5	8	2	4	3	7	6
4	8	7	3	6	9	1	5	2
7	1	3	2	4	5	6	8	9
8	2	4	6	9	3	5	1	7
9	5	6	7	1	8	2	4	3

Solution # 830

6	4	3	9	8	2	5	7	1
5	1	2	7	4	6	3	8	9
8	9	7	3	5	1	2	4	6
9	5	8	4	2	7	1	6	3
3	7	4	6	1	5	8	9	2
1	2	6	8	9	3	4	5	7
2	6	1	5	7	4	9	3	8
7	8	5	2	3	9	6	1	4
4	3	9	1	6	8	7	2	5

Solution # 831

5	8	1	9	3	4	7	6	2
4	3	7	5	6	2	1	8	9
6	2	9	7	1	8	5	4	3
9	5	3	6	4	1	2	7	8
2	7	6	3	8	9	4	1	5
1	4	8	2	7	5	9	3	6
3	6	5	1	2	7	8	9	4
8	1	2	4	9	3	6	5	7
7	9	4	8	5	6	3	2	1

Solution # 832

7	3	6	4	2	8	1	5	9
4	5	8	3	1	9	7	2	6
9	2	1	6	7	5	3	4	8
6	7	9	8	5	2	4	3	1
3	4	5	9	6	1	8	7	2
8	1	2	7	4	3	9	6	5
1	9	4	2	3	6	5	8	7
2	8	3	5	9	7	6	1	4
5	6	7	1	8	4	2	9	3

Solution # 833

6	8	1	7	3	9	4	5	2
5	7	3	2	4	8	1	9	6
2	4	9	6	1	5	7	8	3
8	6	4	3	7	2	9	1	5
7	3	2	5	9	1	6	4	8
9	1	5	4	8	6	2	3	7
4	9	6	8	2	3	5	7	1
3	5	7	1	6	4	8	2	9
1	2	8	9	5	7	3	6	4

Solution # 834

2	4	8	7	6	5	3	9	1
9	5	7	3	8	1	4	2	6
3	6	1	2	4	9	8	5	7
6	9	5	8	2	7	1	3	4
8	2	3	5	1	4	6	7	9
7	1	4	6	9	3	5	8	2
1	8	9	4	5	2	7	6	3
5	7	2	1	3	6	9	4	8
4	3	6	9	7	8	2	1	5

Solution # 835

7	6	9	4	3	5	8	2	1
1	8	4	7	6	2	5	3	9
5	3	2	8	1	9	7	6	4
2	9	6	3	5	7	1	4	8
4	7	3	1	9	8	2	5	6
8	1	5	6	2	4	9	7	3
9	2	8	5	4	6	3	1	7
6	5	1	9	7	3	4	8	2
3	4	7	2	8	1	6	9	5

Solution # 836

5	2	1	7	9	4	6	3	8
6	9	4	5	3	8	7	2	1
3	7	8	1	2	6	9	5	4
8	4	5	2	6	9	3	1	7
9	1	3	8	5	7	4	6	2
2	6	7	4	1	3	5	8	9
1	5	6	9	7	2	8	4	3
7	8	2	3	4	5	1	9	6
4	3	9	6	8	1	2	7	5

Solution # 837

9	6	8	2	5	3	1	7	4
5	7	3	6	4	1	2	8	9
1	2	4	7	9	8	3	6	5
4	9	5	3	7	2	6	1	8
3	1	7	5	8	6	4	9	2
6	8	2	4	1	9	5	3	7
2	3	9	8	6	5	7	4	1
7	5	1	9	3	4	8	2	6
8	4	6	1	2	7	9	5	3

Solution # 838

9	6	2	7	5	3	4	1	8
1	7	8	9	4	6	5	2	3
4	5	3	1	8	2	7	6	9
6	4	9	3	7	1	8	5	2
7	2	1	5	6	8	9	3	4
3	8	5	4	2	9	6	7	1
5	9	4	2	3	7	1	8	6
8	3	7	6	1	4	2	9	5
2	1	6	8	9	5	3	4	7

Solution # 839

9	2	6	4	8	5	1	3	7
8	4	1	7	2	3	9	6	5
7	3	5	6	1	9	8	2	4
1	7	2	3	9	6	5	4	8
5	8	9	2	7	4	3	1	6
4	6	3	8	5	1	7	9	2
3	5	7	1	6	2	4	8	9
2	1	8	9	4	7	6	5	3
6	9	4	5	3	8	2	7	1

Solution # 840

9	4	1	2	7	6	3	5	8
6	5	3	8	1	9	7	4	2
2	7	8	4	3	5	9	6	1
3	8	6	9	2	4	1	7	5
5	2	9	7	6	1	4	8	3
7	1	4	3	5	8	2	9	6
4	9	2	6	8	3	5	1	7
8	3	5	1	4	7	6	2	9
1	6	7	5	9	2	8	3	4

Solution # 841

8	4	3	7	6	5	9	2	1
5	7	9	2	4	1	6	8	3
2	6	1	3	9	8	5	4	7
9	8	2	5	3	4	1	7	6
4	5	7	6	1	2	8	3	9
3	1	6	8	7	9	4	5	2
6	2	8	1	5	7	3	9	4
7	3	4	9	8	6	2	1	5
1	9	5	4	2	3	7	6	8

Solution # 842

2	7	5	3	9	1	4	8	6
4	6	9	2	5	8	7	3	1
1	8	3	7	6	4	5	9	2
8	2	6	5	4	9	1	7	3
7	3	1	6	8	2	9	4	5
9	5	4	1	3	7	6	2	8
3	4	8	9	1	5	2	6	7
6	1	7	4	2	3	8	5	9
5	9	2	8	7	6	3	1	4

Solution # 843

4	8	5	3	9	2	7	1	6
7	6	2	5	8	1	3	9	4
3	1	9	4	6	7	5	8	2
9	3	1	6	5	8	4	2	7
6	7	4	2	3	9	8	5	1
5	2	8	7	1	4	9	6	3
8	4	3	1	2	5	6	7	9
2	5	6	9	7	3	1	4	8
1	9	7	8	4	6	2	3	5

Solution # 844

4	7	8	1	5	3	2	9	6
9	5	2	8	6	7	1	3	4
3	1	6	4	9	2	7	8	5
2	8	7	5	1	9	4	6	3
1	9	3	7	4	6	8	5	2
5	6	4	2	3	8	9	1	7
8	3	9	6	7	4	5	2	1
6	4	1	9	2	5	3	7	8
7	2	5	3	8	1	6	4	9

Solution # 845

5	1	6	3	7	4	2	9	8
7	4	8	2	1	9	3	5	6
3	9	2	8	5	6	1	4	7
1	5	7	4	3	8	9	6	2
6	8	3	9	2	1	4	7	5
4	2	9	7	6	5	8	3	1
2	6	5	1	4	3	7	8	9
9	7	4	6	8	2	5	1	3
8	3	1	5	9	7	6	2	4

Solution # 846

1	4	9	2	3	5	6	8	7
7	2	3	8	4	6	1	5	9
5	8	6	1	9	7	3	2	4
6	1	7	3	8	4	5	9	2
9	5	8	6	1	2	4	7	3
2	3	4	7	5	9	8	6	1
4	6	2	5	7	1	9	3	8
8	9	5	4	2	3	7	1	6
3	7	1	9	6	8	2	4	5

Solution # 847

1	6	2	7	8	4	9	3	5
4	3	7	5	9	6	2	1	8
5	9	8	1	3	2	7	6	4
7	1	6	4	5	9	3	8	2
8	4	5	3	2	7	6	9	1
9	2	3	8	6	1	5	4	7
3	5	1	9	7	8	4	2	6
2	7	4	6	1	3	8	5	9
6	8	9	2	4	5	1	7	3

Solution # 848

3	4	5	7	9	8	1	2	6
9	8	6	2	1	3	4	5	7
2	7	1	5	4	6	8	3	9
7	1	3	9	6	2	5	4	8
6	5	2	8	7	4	9	1	3
8	9	4	1	3	5	7	6	2
4	2	7	6	8	1	3	9	5
1	6	8	3	5	9	2	7	4
5	3	9	4	2	7	6	8	1

Solution # 849

5	2	4	8	6	3	9	7	1
6	9	3	4	1	7	2	8	5
8	1	7	5	2	9	4	3	6
7	8	9	3	4	6	1	5	2
1	5	6	2	7	8	3	9	4
3	4	2	9	5	1	7	6	8
2	6	1	7	9	5	8	4	3
9	3	5	1	8	4	6	2	7
4	7	8	6	3	2	5	1	9

Solution # 850

5	1	2	6	3	9	4	8	7
4	3	6	8	7	5	2	9	1
8	7	9	2	4	1	5	6	3
9	2	8	7	1	4	6	3	5
1	4	5	3	6	8	9	7	2
7	6	3	9	5	2	8	1	4
2	8	4	1	9	7	3	5	6
6	9	7	5	2	3	1	4	8
3	5	1	4	8	6	7	2	9

Solution # 851

1	9	2	6	8	7	4	3	5
7	4	8	1	3	5	2	9	6
5	6	3	9	4	2	7	1	8
9	3	5	8	2	6	1	7	4
4	7	6	3	1	9	5	8	2
2	8	1	7	5	4	9	6	3
3	2	7	4	6	1	8	5	9
8	5	9	2	7	3	6	4	1
6	1	4	5	9	8	3	2	7

Solution # 852

4	7	8	2	6	9	5	1	3
6	9	5	1	4	3	8	7	2
3	1	2	7	8	5	9	6	4
5	4	7	9	1	8	2	3	6
2	3	9	5	7	6	4	8	1
1	8	6	4	3	2	7	9	5
7	2	4	3	9	1	6	5	8
8	5	3	6	2	7	1	4	9
9	6	1	8	5	4	3	2	7

Solution # 853

2	3	6	5	7	4	9	1	8
7	8	9	1	3	6	2	4	5
4	1	5	9	2	8	7	6	3
5	2	3	6	4	9	8	7	1
8	4	7	3	1	5	6	2	9
6	9	1	2	8	7	5	3	4
3	6	8	4	5	2	1	9	7
9	5	4	7	6	1	3	8	2
1	7	2	8	9	3	4	5	6

Solution # 854

3	6	4	1	5	9	7	2	8
2	5	8	6	4	7	3	1	9
9	7	1	8	2	3	4	5	6
7	8	5	9	3	4	2	6	1
6	3	9	2	7	1	5	8	4
4	1	2	5	8	6	9	7	3
8	9	7	3	6	2	1	4	5
1	4	6	7	9	5	8	3	2
5	2	3	4	1	8	6	9	7

Solution # 855

5	8	1	3	9	7	2	4	6
3	4	7	6	8	2	5	1	9
6	9	2	4	1	5	7	3	8
8	6	4	2	7	3	1	9	5
7	5	9	8	6	1	4	2	3
1	2	3	9	5	4	8	6	7
2	3	5	7	4	6	9	8	1
4	1	8	5	3	9	6	7	2
9	7	6	1	2	8	3	5	4

Solution # 856

9	5	1	2	7	3	8	4	6
7	2	6	8	4	5	9	3	1
4	8	3	6	1	9	2	5	7
6	4	8	9	3	7	1	2	5
3	9	2	5	8	1	7	6	4
1	7	5	4	6	2	3	8	9
2	6	4	1	9	8	5	7	3
5	3	9	7	2	6	4	1	8
8	1	7	3	5	4	6	9	2

Solution # 857

8	2	4	7	3	1	6	5	9
5	3	9	6	4	8	2	7	1
7	1	6	5	9	2	4	3	8
4	7	3	2	5	9	1	8	6
2	8	5	3	1	6	7	9	4
6	9	1	8	7	4	5	2	3
3	5	8	4	6	7	9	1	2
1	6	2	9	8	5	3	4	7
9	4	7	1	2	3	8	6	5

Solution # 858

8	5	2	3	6	4	7	9	1
4	1	9	2	5	7	6	8	3
3	7	6	8	9	1	4	2	5
2	8	1	9	7	3	5	4	6
6	3	7	4	2	5	8	1	9
5	9	4	6	1	8	3	7	2
1	6	8	5	4	2	9	3	7
9	2	3	7	8	6	1	5	4
7	4	5	1	3	9	2	6	8

Solution # 859

4	9	3	1	2	5	8	6	7
6	1	2	4	8	7	5	3	9
5	7	8	6	3	9	4	2	1
8	5	1	9	4	6	2	7	3
2	3	9	7	1	8	6	5	4
7	6	4	3	5	2	9	1	8
3	4	5	2	9	1	7	8	6
9	8	6	5	7	3	1	4	2
1	2	7	8	6	4	3	9	5

Solution # 860

2	4	3	9	7	1	8	6	5
1	7	9	5	8	6	2	3	4
8	6	5	4	2	3	9	1	7
5	2	6	3	1	7	4	9	8
7	1	4	6	9	8	5	2	3
9	3	8	2	5	4	6	7	1
6	5	7	8	3	2	1	4	9
3	8	2	1	4	9	7	5	6
4	9	1	7	6	5	3	8	2

Solution # 861

2	6	8	5	9	3	7	1	4
9	4	7	2	1	6	8	3	5
5	3	1	8	4	7	2	6	9
8	5	3	6	7	2	4	9	1
4	2	9	3	5	1	6	7	8
1	7	6	4	8	9	3	5	2
3	8	4	1	6	5	9	2	7
6	9	5	7	2	4	1	8	3
7	1	2	9	3	8	5	4	6

Solution # 862

4	2	1	5	8	6	7	9	3
7	3	8	9	2	4	1	6	5
5	6	9	1	3	7	4	2	8
8	7	4	2	9	3	5	1	6
2	5	3	6	1	8	9	7	4
9	1	6	4	7	5	8	3	2
3	4	5	7	6	9	2	8	1
6	9	2	8	4	1	3	5	7
1	8	7	3	5	2	6	4	9

Solution # 863

4	8	1	2	6	5	7	9	3
7	9	6	3	4	8	1	2	5
2	3	5	9	1	7	4	6	8
1	2	3	6	8	9	5	7	4
8	4	7	5	2	1	9	3	6
6	5	9	4	7	3	2	8	1
5	1	8	7	3	2	6	4	9
3	6	2	1	9	4	8	5	7
9	7	4	8	5	6	3	1	2

Solution # 864

6	3	1	2	5	9	8	4	7
5	2	7	6	4	8	9	3	1
4	8	9	7	3	1	2	6	5
9	7	8	4	1	2	6	5	3
3	1	4	5	9	6	7	2	8
2	5	6	3	8	7	4	1	9
1	6	2	9	7	3	5	8	4
8	9	5	1	2	4	3	7	6
7	4	3	8	6	5	1	9	2

Solution # 865

6	9	5	8	7	2	4	3	1
2	8	3	5	1	4	6	9	7
4	1	7	6	3	9	5	2	8
7	4	6	1	2	8	3	5	9
8	2	1	3	9	5	7	4	6
3	5	9	7	4	6	8	1	2
1	3	4	2	6	7	9	8	5
5	6	2	9	8	3	1	7	4
9	7	8	4	5	1	2	6	3

Solution # 866

1	3	6	8	9	2	5	7	4
9	5	2	4	7	6	1	8	3
7	8	4	1	5	3	6	9	2
3	1	7	9	4	8	2	5	6
5	4	9	6	2	1	8	3	7
6	2	8	7	3	5	4	1	9
4	7	5	2	1	9	3	6	8
8	9	1	3	6	4	7	2	5
2	6	3	5	8	7	9	4	1

Solution # 867

3	2	8	6	1	7	5	9	4
1	5	4	8	2	9	6	7	3
6	7	9	3	4	5	2	8	1
5	6	3	2	7	4	8	1	9
2	8	1	9	5	3	4	6	7
4	9	7	1	6	8	3	5	2
7	3	2	5	8	1	9	4	6
8	4	6	7	9	2	1	3	5
9	1	5	4	3	6	7	2	8

Solution # 868

1	7	3	9	4	5	8	6	2
5	8	6	2	7	1	3	9	4
4	2	9	3	8	6	7	5	1
9	5	1	4	3	8	6	2	7
8	4	7	6	2	9	5	1	3
3	6	2	5	1	7	9	4	8
2	1	5	7	9	3	4	8	6
7	9	8	1	6	4	2	3	5
6	3	4	8	5	2	1	7	9

Solution # 869

3	4	5	2	8	9	1	7	6
1	7	8	5	4	6	2	9	3
6	9	2	3	7	1	4	5	8
4	3	6	9	5	2	8	1	7
9	8	7	4	1	3	5	6	2
2	5	1	8	6	7	3	4	9
5	2	4	6	9	8	7	3	1
8	1	9	7	3	4	6	2	5
7	6	3	1	2	5	9	8	4

Solution # 870

7	9	1	2	4	8	3	5	6
3	6	2	9	1	5	4	7	8
4	8	5	3	6	7	9	2	1
6	4	3	5	2	1	7	8	9
8	5	9	4	7	6	2	1	3
1	2	7	8	3	9	5	6	4
9	1	8	7	5	3	6	4	2
5	3	4	6	8	2	1	9	7
2	7	6	1	9	4	8	3	5

Solution # 871

6	8	7	5	9	4	3	1	2
2	1	9	7	3	8	4	5	6
5	3	4	2	1	6	7	9	8
1	7	6	3	5	2	8	4	9
3	4	5	1	8	9	6	2	7
8	9	2	6	4	7	1	3	5
4	6	3	9	7	5	2	8	1
9	2	8	4	6	1	5	7	3
7	5	1	8	2	3	9	6	4

Solution # 872

1	4	7	5	3	8	6	2	9
9	2	5	4	6	1	3	8	7
8	6	3	7	2	9	1	5	4
3	7	2	9	5	6	4	1	8
4	8	1	2	7	3	5	9	6
6	5	9	8	1	4	2	7	3
7	3	4	1	9	5	8	6	2
5	9	8	6	4	2	7	3	1
2	1	6	3	8	7	9	4	5

Solution # 873

5	7	9	6	4	1	8	2	3
3	1	8	7	2	9	5	4	6
6	4	2	3	8	5	9	1	7
4	3	5	9	1	2	6	7	8
2	8	7	5	3	6	4	9	1
9	6	1	4	7	8	2	3	5
7	9	3	8	5	4	1	6	2
1	5	4	2	6	7	3	8	9
8	2	6	1	9	3	7	5	4

Solution # 874

8	9	6	5	2	3	1	4	7
4	1	5	7	6	8	3	2	9
7	3	2	9	1	4	6	8	5
6	8	1	2	9	5	4	7	3
2	4	3	6	8	7	5	9	1
5	7	9	3	4	1	8	6	2
1	6	7	8	5	9	2	3	4
3	5	8	4	7	2	9	1	6
9	2	4	1	3	6	7	5	8

Solution # 875

5	1	9	3	7	8	6	4	2
4	8	7	2	6	1	9	3	5
2	3	6	5	9	4	7	8	1
7	5	4	6	8	3	1	2	9
3	2	8	9	1	7	5	6	4
6	9	1	4	5	2	8	7	3
8	4	5	7	2	9	3	1	6
9	7	2	1	3	6	4	5	8
1	6	3	8	4	5	2	9	7

Solution # 876

4	7	6	9	2	3	1	5	8
9	2	8	4	5	1	7	3	6
3	5	1	8	7	6	4	2	9
8	9	4	5	1	7	3	6	2
1	6	2	3	8	9	5	7	4
5	3	7	6	4	2	8	9	1
7	1	5	2	6	8	9	4	3
6	8	3	7	9	4	2	1	5
2	4	9	1	3	5	6	8	7

Solution # 877

2	3	4	6	7	9	1	8	5
9	6	8	5	4	1	2	7	3
5	7	1	2	3	8	6	4	9
8	5	6	3	2	7	9	1	4
3	2	9	1	5	4	7	6	8
1	4	7	9	8	6	5	3	2
4	1	5	7	9	3	8	2	6
6	8	2	4	1	5	3	9	7
7	9	3	8	6	2	4	5	1

Solution # 878

9	4	1	3	5	6	2	8	7
8	7	6	1	9	2	3	4	5
3	5	2	4	8	7	6	9	1
6	3	7	5	4	8	1	2	9
5	2	9	6	1	3	4	7	8
4	1	8	2	7	9	5	3	6
7	6	4	9	2	5	8	1	3
2	9	5	8	3	1	7	6	4
1	8	3	7	6	4	9	5	2

Solution # 879

2	9	4	1	7	8	6	5	3
3	8	6	5	2	4	1	9	7
5	1	7	3	9	6	4	8	2
4	3	9	2	5	7	8	1	6
6	5	2	8	1	9	7	3	4
8	7	1	4	6	3	5	2	9
7	2	5	9	4	1	3	6	8
9	6	8	7	3	5	2	4	1
1	4	3	6	8	2	9	7	5

Solution # 880

3	4	5	7	6	2	1	9	8
8	9	7	5	3	1	6	4	2
6	1	2	4	8	9	3	5	7
2	6	9	1	4	7	8	3	5
4	7	8	2	5	3	9	6	1
5	3	1	8	9	6	2	7	4
1	8	3	9	7	4	5	2	6
9	5	4	6	2	8	7	1	3
7	2	6	3	1	5	4	8	9

Solution # 881

5	2	8	9	6	4	7	1	3
3	4	7	1	8	2	9	5	6
9	1	6	5	3	7	2	8	4
6	7	5	4	2	3	1	9	8
1	3	4	8	9	5	6	2	7
2	8	9	7	1	6	3	4	5
4	9	3	6	5	1	8	7	2
7	6	1	2	4	8	5	3	9
8	5	2	3	7	9	4	6	1

Solution # 882

8	4	9	7	1	2	6	5	3
7	1	6	8	5	3	4	9	2
5	2	3	4	6	9	1	7	8
4	5	7	3	8	6	2	1	9
1	9	8	2	4	7	3	6	5
6	3	2	1	9	5	8	4	7
2	8	4	5	7	1	9	3	6
3	6	5	9	2	4	7	8	1
9	7	1	6	3	8	5	2	4

Solution # 883

3	6	4	2	8	7	9	1	5
8	7	2	1	5	9	4	3	6
9	5	1	4	3	6	8	7	2
5	9	8	7	4	1	6	2	3
7	4	3	6	2	8	1	5	9
1	2	6	5	9	3	7	8	4
2	8	5	9	1	4	3	6	7
6	1	9	3	7	5	2	4	8
4	3	7	8	6	2	5	9	1

Solution # 884

1	8	2	7	4	9	3	5	6
3	4	6	5	1	2	7	8	9
7	9	5	3	8	6	2	4	1
8	6	3	2	5	1	4	9	7
9	1	7	8	3	4	5	6	2
5	2	4	6	9	7	8	1	3
4	3	9	1	7	8	6	2	5
2	7	8	9	6	5	1	3	4
6	5	1	4	2	3	9	7	8

Solution # 885

4	9	3	7	1	8	5	2	6
5	6	1	2	4	9	7	8	3
7	2	8	6	3	5	1	9	4
6	5	2	3	9	1	4	7	8
1	4	7	5	8	2	3	6	9
8	3	9	4	6	7	2	5	1
9	8	5	1	2	3	6	4	7
3	7	6	9	5	4	8	1	2
2	1	4	8	7	6	9	3	5

Solution # 886

1	3	2	7	6	5	8	4	9
7	5	8	1	9	4	6	3	2
9	4	6	2	3	8	1	5	7
4	2	1	3	8	7	5	9	6
3	8	9	6	5	1	7	2	4
5	6	7	9	4	2	3	1	8
6	1	4	8	2	3	9	7	5
2	9	3	5	7	6	4	8	1
8	7	5	4	1	9	2	6	3

Solution # 887

9	7	8	2	3	6	4	5	1
2	1	3	7	4	5	6	9	8
4	6	5	1	8	9	3	7	2
3	8	2	9	7	4	5	1	6
1	5	9	8	6	2	7	4	3
7	4	6	3	5	1	2	8	9
5	9	7	6	2	8	1	3	4
8	2	4	5	1	3	9	6	7
6	3	1	4	9	7	8	2	5

Solution # 888

4	7	9	5	2	1	8	3	6
2	8	1	3	4	6	9	5	7
5	3	6	8	9	7	2	1	4
6	5	4	1	3	8	7	9	2
7	1	2	6	5	9	4	8	3
3	9	8	4	7	2	5	6	1
8	2	5	7	6	3	1	4	9
1	6	7	9	8	4	3	2	5
9	4	3	2	1	5	6	7	8

Solution # 889

2	4	1	9	5	6	7	3	8
3	5	8	7	4	2	6	1	9
9	7	6	8	3	1	4	5	2
1	3	2	6	8	4	5	9	7
4	8	7	5	1	9	3	2	6
5	6	9	2	7	3	8	4	1
8	2	3	1	6	5	9	7	4
6	1	5	4	9	7	2	8	3
7	9	4	3	2	8	1	6	5

Solution # 890

4	7	1	5	2	9	3	6	8
5	2	6	7	8	3	4	9	1
3	9	8	4	6	1	5	2	7
9	1	2	3	5	6	7	8	4
7	5	3	2	4	8	9	1	6
6	8	4	1	9	7	2	5	3
1	3	9	6	7	5	8	4	2
8	4	7	9	1	2	6	3	5
2	6	5	8	3	4	1	7	9

Solution # 891

8	1	9	5	4	2	7	6	3
7	5	2	9	3	6	1	4	8
4	6	3	8	7	1	2	9	5
9	8	4	7	1	5	6	3	2
2	7	6	3	8	9	5	1	4
1	3	5	6	2	4	9	8	7
3	9	7	2	6	8	4	5	1
5	2	1	4	9	3	8	7	6
6	4	8	1	5	7	3	2	9

Solution # 892

8	3	2	9	6	1	4	5	7
4	6	7	3	8	5	9	2	1
1	9	5	4	2	7	3	6	8
3	4	8	5	9	2	7	1	6
6	5	1	7	4	3	8	9	2
2	7	9	8	1	6	5	4	3
9	2	3	6	7	4	1	8	5
5	8	6	1	3	9	2	7	4
7	1	4	2	5	8	6	3	9

Solution # 893

8	4	3	1	7	2	9	6	5
1	2	6	5	9	8	4	7	3
5	9	7	3	4	6	8	2	1
4	8	9	6	5	3	2	1	7
3	5	1	7	2	4	6	8	9
6	7	2	8	1	9	5	3	4
7	6	8	9	3	5	1	4	2
9	3	4	2	6	1	7	5	8
2	1	5	4	8	7	3	9	6

Solution # 894

7	5	4	8	3	6	9	1	2
1	2	3	9	7	5	6	8	4
8	6	9	4	2	1	7	5	3
9	1	7	5	8	4	2	3	6
3	8	6	7	9	2	5	4	1
5	4	2	6	1	3	8	7	9
6	9	8	1	4	7	3	2	5
2	7	1	3	5	9	4	6	8
4	3	5	2	6	8	1	9	7

Solution # 895

3	8	2	6	4	9	1	7	5
4	5	7	3	8	1	2	6	9
1	9	6	5	7	2	3	8	4
6	3	4	7	9	5	8	1	2
9	7	1	2	3	8	4	5	6
8	2	5	4	1	6	9	3	7
7	4	8	9	6	3	5	2	1
5	6	3	1	2	4	7	9	8
2	1	9	8	5	7	6	4	3

Solution # 896

7	1	4	5	3	8	6	9	2
2	6	8	9	4	1	7	3	5
3	9	5	2	7	6	8	1	4
6	7	2	1	9	3	5	4	8
8	3	9	4	5	2	1	6	7
5	4	1	8	6	7	9	2	3
4	5	7	6	2	9	3	8	1
9	8	3	7	1	4	2	5	6
1	2	6	3	8	5	4	7	9

Solution # 897

4	1	9	3	8	5	2	6	7
5	6	3	7	2	4	1	9	8
7	8	2	1	6	9	5	3	4
1	7	6	2	5	8	3	4	9
2	5	4	9	7	3	6	8	1
3	9	8	6	4	1	7	5	2
8	2	5	4	3	7	9	1	6
6	3	1	8	9	2	4	7	5
9	4	7	5	1	6	8	2	3

Solution # 898

3	8	2	9	7	1	5	4	6
4	7	9	5	8	6	3	2	1
6	5	1	3	2	4	7	9	8
5	1	3	7	4	9	8	6	2
8	2	4	6	5	3	9	1	7
7	9	6	8	1	2	4	3	5
2	4	8	1	9	5	6	7	3
9	6	7	2	3	8	1	5	4
1	3	5	4	6	7	2	8	9

Solution # 899

5	7	9	8	3	6	1	2	4
6	4	3	1	2	7	9	8	5
2	1	8	4	9	5	6	3	7
9	2	6	7	4	3	5	1	8
4	8	5	2	6	1	7	9	3
7	3	1	9	5	8	4	6	2
3	6	4	5	8	9	2	7	1
8	5	7	6	1	2	3	4	9
1	9	2	3	7	4	8	5	6

Solution # 900

6	7	9	5	1	2	8	4	3
5	2	3	4	8	6	9	1	7
4	1	8	3	9	7	6	5	2
2	6	7	1	5	9	3	8	4
3	5	4	7	6	8	1	2	9
8	9	1	2	3	4	5	7	6
9	3	2	8	7	1	4	6	5
1	4	5	6	2	3	7	9	8
7	8	6	9	4	5	2	3	1

Solution # 901

1	5	9	7	2	3	4	6	8
6	8	3	4	5	1	9	2	7
2	4	7	9	6	8	1	5	3
5	7	4	2	1	6	3	8	9
8	3	1	5	9	7	2	4	6
9	2	6	8	3	4	7	1	5
3	9	2	6	4	5	8	7	1
7	1	5	3	8	2	6	9	4
4	6	8	1	7	9	5	3	2

Solution # 902

7	1	9	5	6	4	3	8	2
5	4	2	9	8	3	7	1	6
8	6	3	7	2	1	4	5	9
9	7	5	8	3	6	2	4	1
1	2	6	4	9	5	8	3	7
4	3	8	1	7	2	9	6	5
3	9	7	6	5	8	1	2	4
2	5	4	3	1	9	6	7	8
6	8	1	2	4	7	5	9	3

Solution # 903

5	4	3	9	8	7	1	2	6
8	9	6	1	5	2	4	3	7
7	1	2	3	4	6	8	9	5
6	5	9	2	7	8	3	1	4
3	2	1	4	6	9	7	5	8
4	8	7	5	3	1	9	6	2
2	6	4	7	1	3	5	8	9
9	3	5	8	2	4	6	7	1
1	7	8	6	9	5	2	4	3

Solution # 904

8	1	6	4	7	9	2	3	5
2	4	3	6	1	5	9	7	8
5	7	9	8	2	3	4	6	1
3	8	1	9	4	6	7	5	2
9	5	7	3	8	2	1	4	6
6	2	4	1	5	7	8	9	3
4	6	2	7	3	8	5	1	9
1	3	8	5	9	4	6	2	7
7	9	5	2	6	1	3	8	4

Solution # 905

4	1	8	6	2	7	9	3	5
5	9	2	4	3	1	6	7	8
3	7	6	8	9	5	4	1	2
1	4	5	7	8	2	3	9	6
8	3	9	5	4	6	1	2	7
2	6	7	9	1	3	5	8	4
6	8	1	3	7	4	2	5	9
7	2	4	1	5	9	8	6	3
9	5	3	2	6	8	7	4	1

Solution # 906

4	1	2	3	5	8	6	7	9
6	3	5	7	2	9	1	8	4
8	9	7	4	6	1	2	5	3
2	7	9	5	1	3	4	6	8
3	4	8	9	7	6	5	1	2
1	5	6	8	4	2	9	3	7
5	6	4	2	8	7	3	9	1
9	8	1	6	3	4	7	2	5
7	2	3	1	9	5	8	4	6

Solution # 907

3	8	9	4	5	2	1	7	6
7	2	5	8	1	6	4	9	3
1	6	4	7	3	9	5	2	8
5	1	6	9	7	3	8	4	2
2	7	8	6	4	5	9	3	1
4	9	3	1	2	8	7	6	5
6	4	7	2	8	1	3	5	9
9	5	1	3	6	7	2	8	4
8	3	2	5	9	4	6	1	7

Solution # 908

6	7	8	4	5	9	3	2	1
9	1	2	6	3	8	5	4	7
5	3	4	7	1	2	8	6	9
7	9	3	2	6	1	4	8	5
1	8	6	9	4	5	7	3	2
4	2	5	3	8	7	9	1	6
8	6	9	1	7	4	2	5	3
3	5	7	8	2	6	1	9	4
2	4	1	5	9	3	6	7	8

Solution # 909

4	1	5	3	8	7	9	6	2
7	2	9	6	1	5	8	4	3
3	8	6	4	9	2	5	1	7
1	7	4	5	3	8	6	2	9
5	9	3	2	6	4	7	8	1
2	6	8	1	7	9	4	3	5
9	3	2	7	4	6	1	5	8
6	5	7	8	2	1	3	9	4
8	4	1	9	5	3	2	7	6

Solution # 910

2	7	1	5	3	9	8	4	6
6	5	3	4	8	1	7	9	2
8	4	9	7	6	2	1	3	5
1	6	4	2	7	5	9	8	3
5	9	2	3	4	8	6	1	7
3	8	7	9	1	6	5	2	4
4	3	5	1	9	7	2	6	8
7	1	6	8	2	4	3	5	9
9	2	8	6	5	3	4	7	1

Solution # 911

2	1	4	9	6	7	3	8	5
9	3	8	1	5	4	6	7	2
7	5	6	3	2	8	9	4	1
4	7	3	5	1	9	8	2	6
6	2	1	8	7	3	4	5	9
8	9	5	2	4	6	1	3	7
1	4	7	6	8	2	5	9	3
5	8	9	7	3	1	2	6	4
3	6	2	4	9	5	7	1	8

Solution # 912

8	3	7	4	6	2	5	9	1
9	4	6	5	1	8	2	7	3
1	5	2	3	9	7	8	6	4
3	7	9	6	4	5	1	8	2
4	2	1	8	3	9	6	5	7
5	6	8	2	7	1	4	3	9
6	9	5	7	2	4	3	1	8
2	1	3	9	8	6	7	4	5
7	8	4	1	5	3	9	2	6

Solution # 913

3	2	6	4	7	9	8	5	1
7	5	4	1	8	6	3	2	9
1	8	9	5	2	3	7	6	4
2	9	1	3	4	7	5	8	6
8	3	5	9	6	2	1	4	7
4	6	7	8	1	5	9	3	2
6	1	2	7	3	8	4	9	5
9	4	8	2	5	1	6	7	3
5	7	3	6	9	4	2	1	8

Solution # 914

5	8	6	7	9	3	2	4	1
2	1	3	6	4	8	9	7	5
7	4	9	2	1	5	6	8	3
4	6	2	5	7	1	3	9	8
8	3	7	9	2	6	1	5	4
1	9	5	8	3	4	7	6	2
3	5	1	4	6	7	8	2	9
6	2	4	3	8	9	5	1	7
9	7	8	1	5	2	4	3	6

Solution # 915

2	1	5	8	7	6	4	9	3
9	7	8	4	5	3	6	1	2
4	3	6	1	9	2	7	5	8
3	2	9	6	8	7	5	4	1
5	6	1	9	3	4	8	2	7
8	4	7	2	1	5	3	6	9
1	5	4	3	2	8	9	7	6
7	8	2	5	6	9	1	3	4
6	9	3	7	4	1	2	8	5

Solution # 916

7	8	6	4	3	9	2	5	1
4	5	3	1	2	7	8	6	9
1	9	2	5	6	8	3	4	7
8	7	5	9	4	2	1	3	6
6	4	1	7	5	3	9	2	8
3	2	9	8	1	6	4	7	5
2	1	4	6	9	5	7	8	3
9	6	7	3	8	4	5	1	2
5	3	8	2	7	1	6	9	4

Solution # 917

8	1	9	5	4	7	3	6	2
5	6	4	3	2	1	7	8	9
3	7	2	9	8	6	1	5	4
1	9	3	4	5	2	6	7	8
2	4	7	8	6	3	9	1	5
6	8	5	1	7	9	2	4	3
4	2	1	6	3	8	5	9	7
7	5	6	2	9	4	8	3	1
9	3	8	7	1	5	4	2	6

Solution # 918

7	3	1	5	2	6	9	4	8
5	6	4	1	9	8	3	7	2
2	9	8	4	7	3	1	5	6
1	2	3	7	5	4	6	8	9
6	4	9	2	8	1	5	3	7
8	5	7	3	6	9	2	1	4
9	7	2	8	3	5	4	6	1
4	8	5	6	1	2	7	9	3
3	1	6	9	4	7	8	2	5

Solution # 919

4	9	2	6	1	5	3	8	7
5	3	6	7	2	8	1	9	4
8	7	1	4	9	3	5	6	2
9	5	4	1	8	6	7	2	3
6	2	3	5	4	7	8	1	9
7	1	8	2	3	9	4	5	6
3	8	5	9	6	4	2	7	1
1	4	9	8	7	2	6	3	5
2	6	7	3	5	1	9	4	8

Solution # 920

3	5	9	2	8	4	7	1	6
1	8	7	9	5	6	3	2	4
2	4	6	3	1	7	5	8	9
7	2	8	5	9	3	6	4	1
5	3	4	1	6	2	8	9	7
9	6	1	4	7	8	2	5	3
8	9	2	7	3	1	4	6	5
6	1	3	8	4	5	9	7	2
4	7	5	6	2	9	1	3	8

Solution # 921

7	4	2	3	6	1	9	8	5
8	5	1	9	4	2	7	6	3
6	3	9	7	8	5	1	2	4
9	1	3	8	5	7	6	4	2
5	6	7	4	2	9	3	1	8
4	2	8	6	1	3	5	9	7
3	8	4	1	7	6	2	5	9
1	9	5	2	3	8	4	7	6
2	7	6	5	9	4	8	3	1

Solution # 922

8	4	3	6	1	5	7	2	9
2	6	7	3	9	4	5	8	1
5	1	9	8	7	2	6	3	4
3	5	4	7	8	1	9	6	2
1	2	8	9	4	6	3	5	7
7	9	6	5	2	3	1	4	8
4	7	5	1	6	8	2	9	3
6	8	1	2	3	9	4	7	5
9	3	2	4	5	7	8	1	6

Solution # 923

7	5	6	8	3	9	2	1	4
4	9	3	5	2	1	6	8	7
8	1	2	7	6	4	5	9	3
3	8	5	6	4	2	9	7	1
2	4	9	1	5	7	8	3	6
1	6	7	9	8	3	4	2	5
5	2	1	3	9	6	7	4	8
9	3	8	4	7	5	1	6	2
6	7	4	2	1	8	3	5	9

Solution # 924

2	4	8	9	5	7	3	6	1
3	7	5	1	6	8	9	4	2
9	6	1	4	3	2	7	8	5
4	1	3	5	8	9	2	7	6
7	5	2	3	4	6	8	1	9
6	8	9	2	7	1	5	3	4
5	3	7	6	2	4	1	9	8
8	9	6	7	1	5	4	2	3
1	2	4	8	9	3	6	5	7

Solution # 925

6	1	5	8	7	4	2	3	9
9	7	3	6	5	2	4	8	1
4	8	2	1	3	9	7	5	6
1	5	4	9	6	8	3	2	7
7	2	9	5	1	3	6	4	8
8	3	6	4	2	7	1	9	5
5	4	1	2	9	6	8	7	3
2	9	7	3	8	1	5	6	4
3	6	8	7	4	5	9	1	2

Solution # 926

9	4	2	6	3	5	1	7	8
6	3	5	7	8	1	9	4	2
8	1	7	4	9	2	6	5	3
3	2	4	9	7	6	8	1	5
1	5	8	3	2	4	7	6	9
7	6	9	5	1	8	3	2	4
4	9	3	1	5	7	2	8	6
2	7	6	8	4	3	5	9	1
5	8	1	2	6	9	4	3	7

Solution # 927

2	5	3	4	6	9	1	7	8
7	8	9	2	5	1	4	3	6
4	1	6	7	3	8	9	5	2
8	6	7	5	2	4	3	9	1
3	9	4	6	1	7	2	8	5
5	2	1	8	9	3	7	6	4
6	4	8	9	7	2	5	1	3
1	7	2	3	8	5	6	4	9
9	3	5	1	4	6	8	2	7

Solution # 928

5	3	1	4	7	9	6	2	8
6	9	2	8	5	3	4	7	1
4	7	8	6	2	1	5	3	9
2	6	7	5	8	4	9	1	3
8	5	3	9	1	2	7	4	6
1	4	9	7	3	6	8	5	2
7	1	6	3	4	8	2	9	5
9	2	4	1	6	5	3	8	7
3	8	5	2	9	7	1	6	4

Solution # 929

3	8	1	7	4	6	9	5	2
9	4	6	5	8	2	7	1	3
7	2	5	9	3	1	4	6	8
5	3	4	1	2	9	8	7	6
8	6	2	3	5	7	1	9	4
1	7	9	4	6	8	3	2	5
2	1	8	6	9	4	5	3	7
4	5	7	2	1	3	6	8	9
6	9	3	8	7	5	2	4	1

Solution # 930

4	3	7	2	1	9	5	8	6
8	9	5	3	7	6	4	2	1
2	1	6	4	5	8	7	3	9
6	7	9	8	2	1	3	4	5
5	4	8	7	6	3	9	1	2
3	2	1	9	4	5	8	6	7
9	5	4	6	8	2	1	7	3
1	8	2	5	3	7	6	9	4
7	6	3	1	9	4	2	5	8

Solution # 931

7	2	8	5	1	4	9	6	3
1	6	3	2	8	9	5	4	7
4	5	9	7	3	6	1	8	2
2	3	5	6	7	8	4	9	1
6	1	7	9	4	3	2	5	8
9	8	4	1	2	5	3	7	6
5	9	1	3	6	7	8	2	4
3	4	6	8	9	2	7	1	5
8	7	2	4	5	1	6	3	9

Solution # 932

9	4	5	3	6	8	7	2	1
2	3	1	5	7	9	4	8	6
7	6	8	2	4	1	3	5	9
5	9	4	7	8	2	1	6	3
1	2	7	6	3	5	9	4	8
6	8	3	1	9	4	5	7	2
4	7	6	9	2	3	8	1	5
8	5	9	4	1	6	2	3	7
3	1	2	8	5	7	6	9	4

Solution # 933

1	2	7	8	6	4	9	5	3
4	9	6	3	2	5	7	8	1
8	3	5	1	9	7	6	4	2
7	8	4	5	1	3	2	6	9
6	1	3	9	4	2	5	7	8
9	5	2	7	8	6	3	1	4
5	6	1	2	3	8	4	9	7
3	4	9	6	7	1	8	2	5
2	7	8	4	5	9	1	3	6

Solution # 934

3	5	7	4	2	6	9	1	8
9	2	1	3	5	8	6	7	4
8	4	6	1	7	9	3	5	2
5	8	4	9	3	2	1	6	7
7	9	3	6	1	4	2	8	5
1	6	2	5	8	7	4	3	9
2	7	9	8	6	3	5	4	1
4	3	5	7	9	1	8	2	6
6	1	8	2	4	5	7	9	3

Solution # 935

3	6	5	2	7	9	8	4	1
8	9	2	3	1	4	6	5	7
4	1	7	8	5	6	2	3	9
2	3	8	5	4	1	7	9	6
5	7	6	9	3	2	1	8	4
1	4	9	6	8	7	5	2	3
7	8	4	1	2	3	9	6	5
9	5	1	4	6	8	3	7	2
6	2	3	7	9	5	4	1	8

Solution # 936

9	2	4	3	7	6	8	5	1
7	6	1	5	8	4	3	9	2
5	3	8	9	2	1	7	4	6
1	4	7	2	3	5	6	8	9
2	5	3	8	6	9	1	7	4
8	9	6	4	1	7	2	3	5
6	8	5	1	4	3	9	2	7
3	7	9	6	5	2	4	1	8
4	1	2	7	9	8	5	6	3

Solution # 937

3	9	7	6	4	8	1	5	2
2	5	6	9	1	3	8	4	7
4	1	8	2	7	5	3	6	9
5	6	3	1	2	9	7	8	4
9	4	2	7	8	6	5	1	3
7	8	1	5	3	4	2	9	6
1	2	9	8	6	7	4	3	5
8	3	5	4	9	2	6	7	1
6	7	4	3	5	1	9	2	8

Solution # 938

6	5	1	3	4	8	2	7	9
2	4	7	5	9	6	3	8	1
3	9	8	1	2	7	4	5	6
4	7	3	9	6	5	8	1	2
5	8	9	2	1	3	7	6	4
1	2	6	8	7	4	9	3	5
7	1	2	6	3	9	5	4	8
9	3	5	4	8	1	6	2	7
8	6	4	7	5	2	1	9	3

Solution # 939

2	7	9	3	5	8	6	1	4
8	1	5	4	7	6	2	3	9
4	3	6	9	2	1	7	5	8
5	9	4	7	1	2	3	8	6
7	8	2	6	4	3	1	9	5
1	6	3	8	9	5	4	2	7
3	4	7	2	8	9	5	6	1
6	5	8	1	3	7	9	4	2
9	2	1	5	6	4	8	7	3

Solution # 940

9	7	8	5	3	2	1	6	4
2	1	5	4	8	6	9	3	7
4	3	6	1	7	9	2	5	8
7	5	1	8	6	3	4	9	2
8	4	9	7	2	5	6	1	3
3	6	2	9	4	1	8	7	5
1	2	7	6	5	8	3	4	9
6	8	4	3	9	7	5	2	1
5	9	3	2	1	4	7	8	6

Solution # 941

6	1	8	2	4	7	3	9	5
2	9	5	1	3	6	8	7	4
3	4	7	8	5	9	2	6	1
5	7	9	3	6	4	1	2	8
4	2	6	9	8	1	7	5	3
1	8	3	5	7	2	9	4	6
7	6	1	4	9	3	5	8	2
8	3	4	7	2	5	6	1	9
9	5	2	6	1	8	4	3	7

Solution # 942

6	1	8	2	9	4	7	5	3
5	2	9	7	8	3	6	1	4
4	3	7	6	5	1	9	8	2
3	6	5	9	1	7	2	4	8
2	7	4	5	6	8	1	3	9
9	8	1	4	3	2	5	7	6
7	5	3	8	2	6	4	9	1
8	4	6	1	7	9	3	2	5
1	9	2	3	4	5	8	6	7

Solution # 943

3	4	9	6	1	7	8	2	5
5	6	7	8	4	2	9	3	1
2	1	8	9	5	3	4	7	6
8	7	4	3	6	1	2	5	9
9	5	6	2	7	8	1	4	3
1	3	2	4	9	5	7	6	8
4	9	3	7	8	6	5	1	2
6	8	1	5	2	4	3	9	7
7	2	5	1	3	9	6	8	4

Solution # 944

3	8	7	6	4	5	1	9	2
5	4	2	9	3	1	7	6	8
6	9	1	7	8	2	4	3	5
1	3	8	2	6	7	5	4	9
2	5	9	3	1	4	8	7	6
4	7	6	8	5	9	3	2	1
7	6	5	1	2	3	9	8	4
8	1	3	4	9	6	2	5	7
9	2	4	5	7	8	6	1	3

Solution # 945

1	9	7	5	8	6	3	4	2
6	5	8	3	4	2	1	9	7
4	2	3	9	7	1	5	8	6
2	7	4	6	1	5	8	3	9
8	6	5	4	9	3	2	7	1
9	3	1	7	2	8	4	6	5
3	8	9	2	5	7	6	1	4
5	4	6	1	3	9	7	2	8
7	1	2	8	6	4	9	5	3

Solution # 946

8	5	6	2	7	9	1	3	4
7	1	2	4	3	5	8	6	9
4	3	9	1	6	8	7	5	2
2	6	5	8	1	7	4	9	3
3	9	8	5	4	6	2	1	7
1	4	7	3	9	2	5	8	6
5	8	4	6	2	3	9	7	1
6	7	1	9	5	4	3	2	8
9	2	3	7	8	1	6	4	5

Solution # 947

3	2	4	5	7	8	6	9	1
5	8	6	1	4	9	2	7	3
9	1	7	2	3	6	8	5	4
1	3	9	8	5	2	4	6	7
4	5	2	6	1	7	3	8	9
7	6	8	4	9	3	5	1	2
6	7	1	3	8	4	9	2	5
2	4	5	9	6	1	7	3	8
8	9	3	7	2	5	1	4	6

Solution # 948

2	8	1	3	4	5	6	7	9
3	6	9	1	7	8	2	4	5
4	5	7	9	6	2	1	3	8
1	9	3	4	8	7	5	2	6
5	4	6	2	1	3	8	9	7
8	7	2	5	9	6	3	1	4
7	1	8	6	2	9	4	5	3
6	3	4	7	5	1	9	8	2
9	2	5	8	3	4	7	6	1

Solution # 949

5	9	8	2	1	6	3	7	4
2	4	7	9	3	8	6	1	5
1	6	3	4	7	5	9	8	2
6	1	4	3	9	7	2	5	8
8	2	9	5	6	1	4	3	7
3	7	5	8	4	2	1	9	6
4	8	6	1	5	3	7	2	9
9	5	1	7	2	4	8	6	3
7	3	2	6	8	9	5	4	1

Solution # 950

4	5	6	3	9	7	1	2	8
3	7	2	4	8	1	5	6	9
1	9	8	2	6	5	3	7	4
2	4	7	5	1	3	9	8	6
8	6	3	7	2	9	4	1	5
5	1	9	6	4	8	2	3	7
7	2	1	8	5	4	6	9	3
9	8	4	1	3	6	7	5	2
6	3	5	9	7	2	8	4	1

Solution # 951

9	1	8	2	5	6	4	7	3
5	2	3	4	7	1	6	9	8
7	4	6	8	3	9	2	1	5
6	5	1	7	8	4	9	3	2
3	8	2	6	9	5	7	4	1
4	9	7	3	1	2	5	8	6
1	6	5	9	4	3	8	2	7
8	3	4	5	2	7	1	6	9
2	7	9	1	6	8	3	5	4

Solution # 952

4	5	2	1	9	6	7	3	8
7	6	3	2	5	8	9	1	4
1	9	8	3	4	7	2	6	5
8	1	4	7	6	3	5	9	2
5	7	6	9	2	1	4	8	3
3	2	9	5	8	4	1	7	6
6	8	1	4	7	2	3	5	9
2	3	5	6	1	9	8	4	7
9	4	7	8	3	5	6	2	1

Solution # 953

9	5	2	3	1	8	6	7	4
1	8	7	4	9	6	2	3	5
4	6	3	5	2	7	1	9	8
2	1	8	9	7	3	5	4	6
3	9	5	6	4	1	7	8	2
6	7	4	2	8	5	9	1	3
5	4	1	8	6	9	3	2	7
7	2	6	1	3	4	8	5	9
8	3	9	7	5	2	4	6	1

Solution # 954

8	2	3	4	6	7	5	1	9
6	9	7	5	1	3	2	4	8
5	4	1	2	8	9	6	7	3
3	8	4	9	5	2	7	6	1
1	6	5	7	3	8	4	9	2
9	7	2	6	4	1	8	3	5
2	3	8	1	7	4	9	5	6
7	1	6	8	9	5	3	2	4
4	5	9	3	2	6	1	8	7

Solution # 955

8	9	4	5	2	3	1	7	6
3	1	2	7	6	9	5	4	8
7	5	6	4	8	1	9	2	3
6	7	8	1	3	2	4	5	9
5	4	9	6	7	8	3	1	2
2	3	1	9	4	5	6	8	7
4	2	3	8	5	6	7	9	1
9	6	5	2	1	7	8	3	4
1	8	7	3	9	4	2	6	5

Solution # 956

7	3	9	8	1	4	5	6	2
4	2	5	9	3	6	7	8	1
8	1	6	2	5	7	9	4	3
6	4	3	7	8	2	1	9	5
9	5	8	1	4	3	2	7	6
1	7	2	5	6	9	8	3	4
3	9	7	4	2	5	6	1	8
5	6	1	3	7	8	4	2	9
2	8	4	6	9	1	3	5	7

Solution # 957

2	8	3	6	7	5	1	9	4
9	4	5	8	3	1	7	6	2
6	7	1	4	2	9	8	5	3
5	9	4	2	1	8	6	3	7
1	6	7	3	5	4	2	8	9
3	2	8	7	9	6	5	4	1
7	5	9	1	6	3	4	2	8
4	1	6	9	8	2	3	7	5
8	3	2	5	4	7	9	1	6

Solution # 958

7	3	1	4	5	9	2	8	6
5	8	6	2	7	1	9	3	4
2	4	9	3	8	6	1	7	5
8	6	2	9	3	7	4	5	1
1	5	4	8	6	2	7	9	3
3	9	7	1	4	5	6	2	8
6	2	3	7	1	8	5	4	9
4	7	5	6	9	3	8	1	2
9	1	8	5	2	4	3	6	7

Solution # 959

4	8	9	2	7	3	5	6	1
6	2	7	1	8	5	4	9	3
5	3	1	4	9	6	2	7	8
3	4	5	9	1	2	7	8	6
7	1	8	5	6	4	9	3	2
2	9	6	7	3	8	1	5	4
1	5	3	8	4	7	6	2	9
9	6	2	3	5	1	8	4	7
8	7	4	6	2	9	3	1	5

Solution # 960

5	6	4	2	3	1	9	8	7
1	3	7	8	5	9	6	2	4
9	2	8	6	4	7	1	3	5
2	9	5	3	1	6	7	4	8
6	8	3	5	7	4	2	9	1
7	4	1	9	8	2	5	6	3
8	1	2	4	6	5	3	7	9
3	5	9	7	2	8	4	1	6
4	7	6	1	9	3	8	5	2

Solution # 961

4	2	8	3	1	7	9	5	6
5	9	1	4	6	8	2	7	3
6	7	3	9	5	2	4	1	8
2	1	5	6	4	3	8	9	7
7	3	6	8	9	1	5	2	4
9	8	4	7	2	5	6	3	1
3	5	2	1	8	4	7	6	9
1	4	9	5	7	6	3	8	2
8	6	7	2	3	9	1	4	5

Solution # 962

5	7	2	3	4	1	9	8	6
3	8	9	2	7	6	4	1	5
4	1	6	8	5	9	3	2	7
6	5	3	9	1	4	8	7	2
9	4	8	7	6	2	1	5	3
7	2	1	5	8	3	6	9	4
8	6	4	1	2	7	5	3	9
2	3	5	4	9	8	7	6	1
1	9	7	6	3	5	2	4	8

Solution # 963

4	8	5	1	2	6	7	3	9
6	2	7	9	4	3	5	8	1
9	1	3	7	8	5	4	2	6
3	9	6	4	5	7	2	1	8
1	7	8	6	3	2	9	4	5
5	4	2	8	9	1	6	7	3
8	6	4	3	7	9	1	5	2
7	5	9	2	1	8	3	6	4
2	3	1	5	6	4	8	9	7

Solution # 964

1	3	7	8	9	4	5	6	2
5	8	9	6	2	1	3	4	7
6	2	4	5	3	7	9	8	1
8	7	5	1	4	9	6	2	3
3	4	6	7	8	2	1	9	5
2	9	1	3	6	5	8	7	4
4	5	2	9	1	6	7	3	8
9	1	8	2	7	3	4	5	6
7	6	3	4	5	8	2	1	9

Solution # 965

9	5	3	7	4	6	8	2	1
4	7	2	8	5	1	6	3	9
6	8	1	2	3	9	5	4	7
7	6	4	9	2	5	3	1	8
5	1	8	3	7	4	9	6	2
3	2	9	6	1	8	4	7	5
2	4	5	1	9	3	7	8	6
8	9	7	4	6	2	1	5	3
1	3	6	5	8	7	2	9	4

Solution # 966

6	9	1	7	2	5	3	4	8
4	5	8	6	1	3	2	7	9
2	3	7	4	9	8	5	1	6
1	2	5	9	8	6	7	3	4
7	4	6	2	3	1	9	8	5
3	8	9	5	4	7	6	2	1
5	6	4	8	7	2	1	9	3
8	7	3	1	6	9	4	5	2
9	1	2	3	5	4	8	6	7

Solution # 967

5	4	8	6	3	7	9	1	2
9	3	1	2	5	8	4	6	7
2	7	6	1	4	9	8	3	5
6	9	4	7	1	3	5	2	8
1	5	3	4	8	2	6	7	9
8	2	7	5	9	6	1	4	3
4	1	9	3	2	5	7	8	6
3	6	5	8	7	4	2	9	1
7	8	2	9	6	1	3	5	4

Solution # 968

6	7	5	2	3	4	9	8	1
8	3	4	1	9	7	2	6	5
2	1	9	6	5	8	4	7	3
3	6	2	4	7	1	8	5	9
5	4	8	3	2	9	6	1	7
7	9	1	5	8	6	3	2	4
1	2	3	8	4	5	7	9	6
4	5	7	9	6	2	1	3	8
9	8	6	7	1	3	5	4	2

Solution # 969

1	4	3	6	8	5	9	2	7
5	8	6	9	7	2	4	1	3
2	9	7	3	1	4	5	8	6
7	2	4	1	5	6	8	3	9
3	5	9	8	2	7	6	4	1
8	6	1	4	3	9	7	5	2
6	1	2	5	9	8	3	7	4
9	3	5	7	4	1	2	6	8
4	7	8	2	6	3	1	9	5

Solution # 970

2	4	6	3	1	8	5	9	7
1	9	3	7	5	6	4	8	2
7	5	8	2	9	4	1	6	3
3	6	2	8	4	9	7	5	1
5	8	7	6	2	1	3	4	9
4	1	9	5	3	7	8	2	6
6	7	1	9	8	5	2	3	4
9	3	5	4	7	2	6	1	8
8	2	4	1	6	3	9	7	5

Solution # 971

5	6	1	8	2	7	9	3	4
2	7	9	3	5	4	1	8	6
8	4	3	1	6	9	7	2	5
1	5	7	4	9	8	3	6	2
3	2	8	6	7	5	4	1	9
6	9	4	2	1	3	8	5	7
4	1	2	7	8	6	5	9	3
7	8	5	9	3	2	6	4	1
9	3	6	5	4	1	2	7	8

Solution # 972

2	5	3	6	8	4	1	9	7
1	7	8	2	9	5	3	4	6
4	9	6	3	7	1	2	8	5
7	1	4	9	5	3	6	2	8
3	8	2	4	1	6	5	7	9
9	6	5	8	2	7	4	3	1
6	2	9	5	3	8	7	1	4
5	3	7	1	4	9	8	6	2
8	4	1	7	6	2	9	5	3

Solution # 973

6	8	7	4	2	9	1	3	5
4	5	1	3	6	8	7	2	9
9	2	3	1	5	7	6	4	8
5	4	6	8	1	3	2	9	7
3	9	2	7	4	6	8	5	1
1	7	8	2	9	5	4	6	3
8	1	9	6	3	2	5	7	4
2	3	4	5	7	1	9	8	6
7	6	5	9	8	4	3	1	2

Solution # 974

8	3	7	1	9	6	2	5	4
4	6	1	8	2	5	7	3	9
5	2	9	4	3	7	8	1	6
6	1	5	7	4	3	9	8	2
3	9	4	2	5	8	6	7	1
2	7	8	6	1	9	5	4	3
7	4	6	3	8	2	1	9	5
1	5	2	9	7	4	3	6	8
9	8	3	5	6	1	4	2	7

Solution # 975

5	2	9	4	6	8	1	7	3
7	6	8	1	3	2	5	4	9
3	4	1	9	7	5	8	2	6
2	3	5	8	9	1	7	6	4
1	9	6	2	4	7	3	8	5
4	8	7	6	5	3	2	9	1
8	5	4	3	2	6	9	1	7
9	7	2	5	1	4	6	3	8
6	1	3	7	8	9	4	5	2

Solution # 976

5	4	7	8	3	6	9	2	1
9	3	6	1	4	2	7	5	8
1	8	2	7	5	9	6	3	4
3	2	5	9	6	1	4	8	7
4	7	1	5	2	8	3	6	9
8	6	9	4	7	3	5	1	2
7	5	8	6	1	4	2	9	3
6	9	3	2	8	7	1	4	5
2	1	4	3	9	5	8	7	6

Solution # 977

4	3	5	6	2	1	8	7	9
8	7	2	4	9	5	3	6	1
1	9	6	3	7	8	5	2	4
3	1	8	7	5	6	4	9	2
2	6	7	9	3	4	1	5	8
9	5	4	1	8	2	6	3	7
6	2	3	8	4	7	9	1	5
5	8	9	2	1	3	7	4	6
7	4	1	5	6	9	2	8	3

Solution # 978

1	7	9	8	3	6	4	5	2
4	6	5	9	1	2	8	7	3
8	2	3	4	7	5	1	6	9
2	4	8	5	9	7	3	1	6
6	9	1	2	4	3	7	8	5
3	5	7	6	8	1	2	9	4
9	3	4	7	5	8	6	2	1
5	8	6	1	2	4	9	3	7
7	1	2	3	6	9	5	4	8

Solution # 979

5	4	1	3	6	7	9	2	8
6	9	2	1	5	8	4	7	3
8	7	3	9	2	4	5	6	1
4	8	5	6	1	9	7	3	2
7	2	6	8	4	3	1	9	5
3	1	9	5	7	2	8	4	6
1	3	8	4	9	6	2	5	7
2	6	4	7	8	5	3	1	9
9	5	7	2	3	1	6	8	4

Solution # 980

9	5	7	8	3	4	6	2	1
4	6	1	5	2	9	7	8	3
3	8	2	6	7	1	4	5	9
2	7	6	3	1	5	9	4	8
5	1	4	7	9	8	3	6	2
8	9	3	2	4	6	5	1	7
1	3	8	4	5	7	2	9	6
6	2	5	9	8	3	1	7	4
7	4	9	1	6	2	8	3	5

Solution # 981

2	6	8	5	4	9	3	7	1
5	3	9	8	7	1	2	4	6
7	4	1	2	3	6	5	9	8
4	7	3	6	5	8	9	1	2
9	2	5	3	1	4	8	6	7
1	8	6	7	9	2	4	5	3
3	9	4	1	2	7	6	8	5
8	5	7	4	6	3	1	2	9
6	1	2	9	8	5	7	3	4

Solution # 982

7	3	5	1	9	2	4	8	6
2	8	9	4	6	5	7	1	3
6	4	1	8	7	3	9	2	5
9	6	7	2	5	1	3	4	8
4	2	8	7	3	6	1	5	9
5	1	3	9	8	4	2	6	7
8	7	4	6	1	9	5	3	2
3	9	2	5	4	8	6	7	1
1	5	6	3	2	7	8	9	4

Solution # 983

4	9	8	2	3	6	5	1	7
6	3	7	5	1	4	2	8	9
5	2	1	7	9	8	3	6	4
2	4	9	1	5	3	8	7	6
1	7	3	6	8	9	4	2	5
8	5	6	4	2	7	9	3	1
3	8	5	9	7	1	6	4	2
9	1	4	3	6	2	7	5	8
7	6	2	8	4	5	1	9	3

Solution # 984

8	7	6	4	1	5	9	2	3
4	5	2	7	9	3	8	6	1
1	3	9	6	8	2	5	7	4
7	9	3	2	4	1	6	8	5
5	4	1	8	7	6	3	9	2
6	2	8	5	3	9	1	4	7
2	1	7	9	5	8	4	3	6
9	6	5	3	2	4	7	1	8
3	8	4	1	6	7	2	5	9

Solution # 985

4	6	9	1	2	8	7	3	5
3	2	5	7	9	4	1	8	6
8	7	1	5	3	6	2	9	4
5	8	2	4	1	9	6	7	3
1	3	4	6	5	7	9	2	8
7	9	6	3	8	2	4	5	1
2	1	7	8	6	3	5	4	9
9	5	3	2	4	1	8	6	7
6	4	8	9	7	5	3	1	2

Solution # 986

4	2	1	6	3	7	5	8	9
8	7	9	2	5	4	6	1	3
3	6	5	9	8	1	2	4	7
6	3	8	7	2	9	1	5	4
7	1	2	5	4	6	9	3	8
5	9	4	3	1	8	7	2	6
2	8	6	1	7	3	4	9	5
1	4	7	8	9	5	3	6	2
9	5	3	4	6	2	8	7	1

Solution # 987

2	1	4	9	7	3	5	8	6
7	6	5	1	4	8	2	3	9
3	8	9	2	5	6	4	1	7
6	7	3	8	2	4	1	9	5
4	5	1	3	6	9	7	2	8
8	9	2	5	1	7	3	6	4
5	3	6	7	8	1	9	4	2
1	4	7	6	9	2	8	5	3
9	2	8	4	3	5	6	7	1

Solution # 988

9	4	1	7	2	6	8	5	3
2	6	8	5	3	1	7	9	4
5	7	3	4	8	9	6	2	1
7	2	6	9	1	3	5	4	8
3	9	4	8	6	5	2	1	7
8	1	5	2	4	7	9	3	6
6	3	2	1	5	8	4	7	9
4	8	7	3	9	2	1	6	5
1	5	9	6	7	4	3	8	2

Solution # 989

1	5	4	7	9	2	3	6	8
2	9	6	8	4	3	7	1	5
3	8	7	5	1	6	2	4	9
9	7	2	4	3	5	1	8	6
5	3	8	1	6	7	4	9	2
4	6	1	9	2	8	5	7	3
8	4	5	3	7	9	6	2	1
6	1	9	2	5	4	8	3	7
7	2	3	6	8	1	9	5	4

Solution # 990

8	7	2	4	9	6	3	5	1
4	5	9	3	1	8	7	6	2
6	3	1	5	2	7	4	9	8
5	2	4	6	3	1	9	8	7
3	1	8	9	7	5	6	2	4
7	9	6	8	4	2	5	1	3
2	4	7	1	5	9	8	3	6
9	6	3	2	8	4	1	7	5
1	8	5	7	6	3	2	4	9

Solution # 991

6	1	5	2	4	7	9	8	3
9	4	8	1	6	3	7	5	2
3	7	2	5	9	8	4	1	6
5	3	9	4	7	2	1	6	8
7	2	1	6	8	5	3	9	4
8	6	4	9	3	1	2	7	5
2	8	6	7	1	4	5	3	9
1	5	3	8	2	9	6	4	7
4	9	7	3	5	6	8	2	1

Solution # 992

2	4	9	1	3	8	6	7	5
1	3	7	4	6	5	2	9	8
6	8	5	9	7	2	1	4	3
4	2	8	5	9	6	3	1	7
9	5	6	3	1	7	4	8	2
3	7	1	8	2	4	5	6	9
7	9	3	6	5	1	8	2	4
5	1	4	2	8	9	7	3	6
8	6	2	7	4	3	9	5	1

Solution # 993

8	1	6	7	3	5	4	9	2
2	4	5	6	8	9	3	1	7
7	3	9	2	1	4	5	6	8
4	9	2	3	7	6	8	5	1
3	6	8	9	5	1	7	2	4
5	7	1	8	4	2	6	3	9
9	5	3	4	2	8	1	7	6
1	2	4	5	6	7	9	8	3
6	8	7	1	9	3	2	4	5

Solution # 994

2	3	1	9	5	4	7	6	8
7	5	9	8	1	6	3	2	4
4	6	8	7	3	2	1	5	9
8	1	4	2	9	3	5	7	6
6	7	3	5	8	1	4	9	2
9	2	5	6	4	7	8	3	1
5	9	2	4	7	8	6	1	3
3	4	6	1	2	5	9	8	7
1	8	7	3	6	9	2	4	5

Solution # 995

1	2	6	3	7	9	4	8	5
4	8	9	2	5	1	3	6	7
7	5	3	4	8	6	2	1	9
3	1	7	5	9	4	6	2	8
2	4	5	6	3	8	9	7	1
6	9	8	7	1	2	5	4	3
8	7	2	9	6	3	1	5	4
9	6	1	8	4	5	7	3	2
5	3	4	1	2	7	8	9	6

Solution # 996

5	8	6	1	9	7	3	2	4
2	4	7	5	3	6	8	9	1
3	9	1	2	8	4	5	7	6
9	5	4	7	6	2	1	8	3
6	2	8	3	1	5	9	4	7
7	1	3	8	4	9	2	6	5
1	7	2	4	5	8	6	3	9
8	6	5	9	7	3	4	1	2
4	3	9	6	2	1	7	5	8

Solution # 997

6	2	8	1	4	9	3	7	5
3	9	7	6	8	5	4	1	2
4	5	1	3	7	2	9	8	6
8	4	5	9	2	3	7	6	1
1	3	9	5	6	7	8	2	4
7	6	2	4	1	8	5	3	9
5	1	6	7	3	4	2	9	8
2	7	4	8	9	6	1	5	3
9	8	3	2	5	1	6	4	7

Solution # 998

2	4	1	8	7	5	9	6	3
9	7	5	3	1	6	8	2	4
6	3	8	4	2	9	1	7	5
3	2	4	5	6	8	7	1	9
1	9	7	2	4	3	6	5	8
5	8	6	1	9	7	4	3	2
7	5	9	6	8	2	3	4	1
4	6	2	9	3	1	5	8	7
8	1	3	7	5	4	2	9	6

Solution # 999

9	4	1	5	7	8	6	3	2
2	7	5	6	3	4	9	8	1
3	8	6	1	2	9	4	5	7
8	5	4	2	9	3	1	7	6
1	2	9	7	5	6	8	4	3
6	3	7	4	8	1	2	9	5
7	9	2	8	6	5	3	1	4
5	1	8	3	4	2	7	6	9
4	6	3	9	1	7	5	2	8

Solution # 1000

8	6	2	5	3	7	1	9	4
7	3	4	9	2	1	8	6	5
5	9	1	8	6	4	3	2	7
9	5	3	4	1	8	2	7	6
6	2	8	7	5	3	9	4	1
4	1	7	2	9	6	5	3	8
3	8	6	1	7	2	4	5	9
1	7	5	3	4	9	6	8	2
2	4	9	6	8	5	7	1	3

Solution # 1001

2	8	9	5	7	6	3	4	1
4	7	1	3	9	8	6	2	5
5	3	6	4	2	1	9	7	8
1	4	3	6	5	7	8	9	2
6	2	8	9	1	4	5	3	7
7	9	5	2	8	3	4	1	6
8	5	2	7	3	9	1	6	4
9	1	4	8	6	2	7	5	3
3	6	7	1	4	5	2	8	9

Solution # 1002

2	3	5	6	7	4	1	9	8
6	4	1	8	9	5	3	7	2
8	7	9	2	1	3	4	6	5
1	6	4	5	2	7	9	8	3
3	5	8	4	6	9	7	2	1
7	9	2	1	3	8	5	4	6
5	1	7	9	8	2	6	3	4
4	2	3	7	5	6	8	1	9
9	8	6	3	4	1	2	5	7

Solution # 1003

9	3	7	8	2	6	4	1	5
5	6	1	3	7	4	9	8	2
2	8	4	9	5	1	7	3	6
4	5	6	1	8	3	2	7	9
8	9	2	4	6	7	1	5	3
7	1	3	5	9	2	6	4	8
3	7	9	2	4	8	5	6	1
1	4	5	6	3	9	8	2	7
6	2	8	7	1	5	3	9	4

Solution # 1004

4	1	2	8	5	6	9	7	3
6	3	5	7	1	9	4	2	8
8	7	9	2	3	4	1	6	5
3	5	8	1	2	7	6	9	4
7	2	6	4	9	8	3	5	1
9	4	1	5	6	3	7	8	2
2	9	3	6	8	1	5	4	7
1	8	7	9	4	5	2	3	6
5	6	4	3	7	2	8	1	9

Solution # 1005

4	7	3	9	6	5	8	2	1
8	6	1	2	7	3	9	5	4
5	2	9	8	1	4	7	6	3
1	4	7	5	9	2	3	8	6
9	8	2	7	3	6	4	1	5
6	3	5	1	4	8	2	7	9
2	1	4	3	5	7	6	9	8
3	9	8	6	2	1	5	4	7
7	5	6	4	8	9	1	3	2

Solution # 1006

9	4	6	3	5	7	2	1	8
1	7	8	9	2	4	3	6	5
3	2	5	1	8	6	4	7	9
5	3	4	2	7	8	6	9	1
8	9	1	6	3	5	7	2	4
2	6	7	4	9	1	5	8	3
7	5	3	8	6	9	1	4	2
6	1	9	5	4	2	8	3	7
4	8	2	7	1	3	9	5	6

Solution # 1007

2	9	5	7	1	3	6	4	8
6	1	7	8	4	5	3	2	9
4	3	8	9	6	2	5	1	7
8	4	9	3	5	1	7	6	2
1	6	2	4	9	7	8	3	5
5	7	3	6	2	8	4	9	1
3	5	4	1	8	9	2	7	6
9	2	6	5	7	4	1	8	3
7	8	1	2	3	6	9	5	4

Solution # 1008

8	2	7	4	9	1	5	3	6
4	3	5	8	2	6	1	7	9
1	9	6	3	5	7	4	8	2
3	5	2	7	4	8	6	9	1
9	6	4	1	3	2	7	5	8
7	8	1	9	6	5	2	4	3
2	4	8	6	7	9	3	1	5
5	7	9	2	1	3	8	6	4
6	1	3	5	8	4	9	2	7

www.ingramcontent.com/pod-product-compliance
Lightning Source LLC
Chambersburg PA
CBHW080827220526
45467CB00008B/2213
* 9 7 9 8 3 9 1 5 5 2 8 9 5 *